Michael Erdmann

29. November 1987

David Freedman

BROWNIAN MOTION AND DIFFUSION

With 38 Figures

Springer-Verlag
New York Heidelberg Berlin

David Freedman
Department of Statistics
University of California
Berkeley, CA 94720
U.S.A.

AMS Subject Classification: 60J60, 60J65, 58G32

Library of Congress Cataloging in Publication Data
Freedman, David, 1938–
Brownian motion and diffusion.
Originally published: San Francisco: Holden-Day, 1971 (Holden-Day series in probability and statistics)
Bibliography: p.
Includes index.
1. Brownian motion processes. 2. Diffusion processes. I. Title. II. Series: Holden-Day series in probability and statistics.
QA274.75.F74 1983 519.2′33 82-19573

The original version of this book was published by Holden-Day, Inc. in 1971.

Printed and bound by R. R. Donnelley & Sons, Harrisonburg, VA.
Printed in the United States of America.

9 8 7 6 5 4 3 2 1

ISBN 0-387-**90805**-6 Springer-Verlag New York Heidelberg Berlin
ISBN 3-540-**90805**-6 Springer-Verlag Berlin Heidelberg New York

TO WILLIAM FELLER

PREFACE

A long time ago I started writing a book about Markov chains, Brownian motion, and diffusion. I soon had two hundred pages of manuscript and my publisher was enthusiastic. Some years and several drafts later, I had a thousand pages of manuscript, and my publisher was less enthusiastic. So we made it a trilogy:

Markov Chains
Brownian Motion and Diffusion
Approximating Countable Markov Chains

familiarly — *MC, B & D,* and *ACM*.

I wrote the first two books for beginning graduate students with some knowledge of probability; if you can follow Sections 3.4 to 3.9 of *Brownian Motion and Diffusion* you're in. The first two books are quite independent of one another, and completely independent of the third. This last book is a monograph, which explains one way to think about chains with instantaneous states. The results in it are supposed to be new, except where there are specific disclaimers; it's written in the framework of *Markov Chains*.

Most of the proofs in the trilogy are new, and I tried hard to make them explicit. The old ones were often elegant, but I seldom saw what made them go. With my own, I can sometimes show you why things work. And, as I will argue in a minute, my demonstrations are easier technically. If I wrote them down well enough, you may come to agree.

The approach in all three books is constructive: I did not use the notion of separability for stochastic processes and in general avoided the uncountable axiom of choice. Separability is a great idea for dealing with any really large class of processes. For Brownian motion I find it less satisfactory. To begin with, a theorem on Brownian motion typically amounts to a statement about a probability on a Borel σ-field. It's a shame to have the proof depend on the existence of an unnamable set. Also, separability proofs usually have two parts. There is an abstract part which establishes the existence of a separable version. And there is a combinatorial argument, which establishes some property of the separable version by looking at the behavior of the process on a countable set of times. If you take the constructive approach, the combinatorial argument alone is enough proof.

When I started writing, I believed in regular conditional distributions. To me they're natural and intuitive objects, and the first draft was full of them. I told it like it was, and if the details were a little hard to supply, that was the reader's problem. Eventually I got tired of writing a book intelligible only to me. And I came to believe that in most proofs, the main point is estimating a probability number: the fewer complicated intermediaries, the better. So I switched to computing integrals by Fubini. This is a more powerful technique than you might think and it makes for proofs that can be checked. Virtually all the conditional distributions were banished to the Appendix. The major exception is Chapter 4 of *Markov Chains,* where the vividness of the conditional distribution language compensates for its technical difficulty.

In *Markov Chains,* Chapters 3 to 6 and 8 cover material not usually available in textbooks — for instance: invariance principles for functionals of a Markov chain; Kolmogorov's inequality on the concentration function; the boundary, with examples; and the construction of a variety of continuous-time chains from their jump processes and holding times. Some of these constructions are part of the folklore, but I think this is the first careful public treatment.

Brownian Motion and Diffusion dispenses with most of the customary transform apparatus, again for the sake of computing probability numbers more directly. The chapter on Brownian motion emphasizes topics which haven't had much textbook coverage, like square variation, the reflection principle, and the invariance principle. The chapter on diffusion shows how to obtain the process from Brownian motion by changing time.

I studied with the great men for a time, and saw what they did. The trilogy is what I learned. All I can add is my recommendation that you buy at least one copy of each book.

User's guide to *Brownian Motion and Diffusion*

Chapter 1 is about Brownian motion, and Chapter 2 is about diffusion. The first nine sections of each chapter cover the basic results and take a semester to do. Sections 1.9-11 and 2.10-15 are excursions; 1.9 and 2.14 just record my personal opinion on two issues without much of an argument; 2.12-13 present two Markov chain examples constructed from Brownian motion. I think you will enjoy Section 1.11 on Wald's identity; 2.15 is supposed to help you understand the strong Markov property.

Sections 2.1-7 are independent of Chapter 1; but 2.8-9 depend on theorems 6, 17, 24 and 114 of Chapter 1.

Sections 3.1-3 explain the cruel and unusual notation, and the reference system; 3.4-9 review probability theory quickly; 3.10-17 do the more exotic analysis which I've found useful at various places in the trilogy; and a few things are in 3.10-17 just because I like them.

Chapter 3 is repeated in *MC* and *ACM*. The three books have a common preface and bibliography. Each has its own index and symbol finder.

Acknowledgments

Much of the trilogy is an exposition of the work of other mathematicians, who sometimes get explicit credit for their ideas. Writing *Markov Chains* would have been impossible without constant reference to Chung (1960). Doob (1953) and Feller (1968) were also heavy involuntary contributors. The diffusion part of *Brownian Motion and Diffusion* is a peasant's version of Itô and McKean (1965).

The influence of David Blackwell, Lester Dubins and Roger Purves will be found on many pages, as will that of my honored teacher, William Feller. Ronald Pyke and Harry Reuter read large parts of the manuscript and made an uncomfortably large number of excellent suggestions, many of which I was forced to accept. I also tested drafts on several generations of graduate students, who were patient, encouraging and helpful. These drafts were faithfully typed from the cuneiform by Gail Salo.

The Sloan Foundation and the US Air Force Office of Scientific Research supported me for various periods, always generously, while I did the writing. I finished two drafts while visiting the Hebrew University in Jerusalem, Imperial College in London, and the University of Tel Aviv. I am grateful to the firm of Cohen, Leithman, Kaufman, Yarosky and Fish, criminal lawyers and xerographers in Montreal. And I am still nostalgic for Cohen's Bar in Jerusalem, the caravansary where I wrote the first final draft of *Approximating Countable Markov Chains*.

David Freedman

Berkeley, California
July, 1970

Preface to the Springer edition

My books on *Markov Chains*, *Brownian Motion and Diffusion*, and *Approximating Countable Markov Chains*, were first published in the early 1970's and have not been readily available since then. However, there still seems to be some substantial interest in them, perhaps due to their constructive and set-theoretic flavor, and the extensive use of concrete examples. I am pleased that Springer-Verlag has agreed to reprint the books, making them available again to the scholarly public. I have taken the occasion to correct many small errors, and to add a few references to new work.

David Freedman

Berkeley, California

September, 1982

TABLE OF CONTENTS

1

BROWNIAN MOTION

1. INTRODUCTION

(1) Definition. Normalized Brownian motion B *is a stochastic process* $\{B(t): 0 \leqq t < \infty\}$ *on a probability triple* $(\Omega, \mathscr{F}, \mathscr{P})$, *with properties* (a–c). *That is,* $B(t, \omega)$ *is a function of nonnegative real numbers* t *and points* $\omega \in \Omega$, *such that* $B(t, \cdot)$ *is* $\mathscr{F}$*-measurable for each* t, *and*:

(a) $B(0, \omega) = 0$ *for each* ω;

(b) $B(\cdot, \omega)$ *is continuous for each* ω;

(c) *for* $0 < t_1 < t_2 < \cdots < t_{n-1} < t_n$, *the increments*

$$B(t_1), B(t_2) - B(t_1), \ldots, B(t_n) - B(t_{n-1})$$

are independent and normally distributed, with means 0 *and variances*

$$t_1, t_2 - t_1, \ldots, t_n - t_{n-1}.$$

I remind you that Z is *normally distributed* with mean 0 and variance 1, or $N(0, 1)$ for short, iff Z has density

$$z \to \frac{1}{\sqrt{2\pi}} e^{-\frac{1}{2}z^2}.$$

And Z is normally distributed with mean μ and variance σ^2, or $N(\mu, \sigma^2)$ for short, iff $(Z - \mu)/\sigma$ is $N(0, 1)$.

One of the leading results on Brownian motion is that it exists. The first successful proof is due to Wiener (1923); a rearrangement (6) of it is given in Section 2. Here is the idea. You can construct $B(t)$ for nonnegative binary rational t on a suitable triple $(\Omega, \mathscr{F}, \mathscr{P})$ so as to satisfy (a), and (c) for binary rational t_m. You then show that for $\mathscr{P}$-almost all ω, the function $B(\cdot, \omega)$ is

---

I want to thank Richard Drogin for a careful reading of this chapter.

uniformly continuous on all finite intervals. For good ω, define the new $B(\cdot, \omega)$ on $[0, \infty)$ as the unique continuous function whose retraction to the binary rationals is the old $B(\cdot, \omega)$. For bad ω, define the new $B(\cdot, \omega)$ as the constant 0. The rest of Section 2 concerns estimates and symmetry properties that will be useful later.

Section 3 opens with the *strong Markov property* (21). Hunt (1956) was the first to state this basic result sharply, and give a proof. Suppose τ is a Markov time and is finite almost surely. Then $B(\tau + \cdot) - B(\tau)$ is again normalized Brownian motion, and is independent of the pre-τ sigma field. This fact will be used to establish the *reflection principle* (25). Keep τ Markov and finite almost surely. Let Y be B reflected at τ:

$$(2) \qquad \begin{aligned} Y(t) &= B(t) && \text{for } t \leqq \tau \\ &= 2B(\tau) - B(t) && \text{for } t \geqq \tau. \end{aligned}$$

Geometrically, Y agrees with B up to time τ; after this time Y is B reflected about the horizontal line at height $B(\tau)$. This is illustrated in Figure 1.

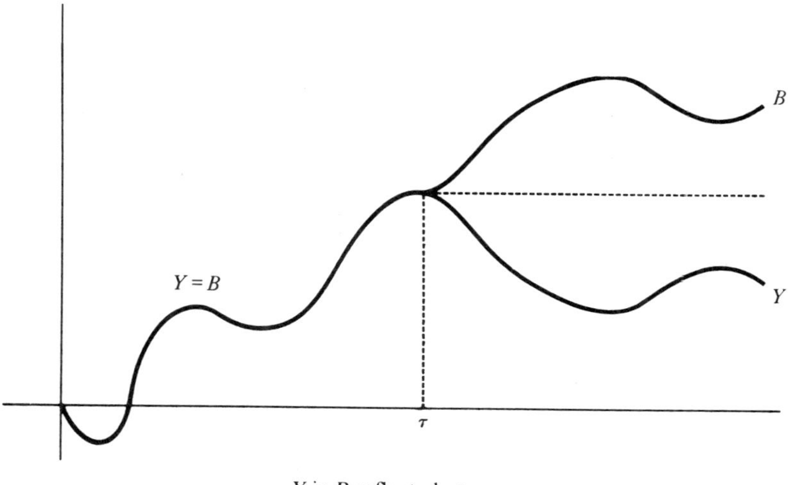

Y is B reflected at τ.

Figure 1.

The reflection principle asserts:

$$Y \quad \text{is also normalized Brownian motion.}$$

This principle will be used to obtain the joint distribution of $B(t)$ and $M(t)$, where

$$M(t) = \max \{B(s) : 0 \leqq s \leqq t\}.$$

Namely (29),

$$\mathscr{P}\{B(t) < m - y \text{ and } M(t) \geqq m\} = \mathscr{P}\{B(t) > m + y\}$$

for nonnegative m and y. This distribution result implies (42): the process

$$\{M(t) - B(t) : 0 \leqq t < \infty\}$$

is distributed like the process

$$\{|B(t)| : 0 \leqq t < \infty\};$$

both are Markov with the same transitions and the same starting state. A byproduct of the argument is that Brownian motion paths are dense. More precisely, let f be a continuous function on $[0, 1]$, with $f(0) = 0$. And let ε be positive. Then (38)

$$\mathscr{P}\{|B(t) - f(t)| \leqq \varepsilon \text{ for } 0 \leqq t \leqq 1\}$$

is positive.

Section 4 deals with sample function properties. Delete a set of exceptional ω, of total probability 0.

(a) All local maxima of $B(\cdot, \omega)$ are strict, by (52).

(b) The local maxima of $B(\cdot, \omega)$ form a countable, dense set, by (51).

(c) $\{t : B(t, \omega) = 0\}$ is closed, dense in itself, and of Lebesgue measure 0; in particular, this set includes no interval, but has the power of the continuum: by (57).

(d) $B(\cdot, \omega)$ is nowhere differentiable. In fact (60), at all times either the upper right derivative of $B(\cdot, \omega)$ is $+\infty$ or the lower right derivative is $-\infty$.

(e) Relative to a fixed sequence of partitions which becomes dense, the square variation of $B(\cdot, \omega)$ on an interval is equal to the length of the interval, by (61).

The exceptional null set in (e) really depends on the partition sequence; its union over all sequences is the whole of Ω.

(f) $B(\cdot, \omega)$ is of bounded variation on no interval, by (71).

This is forced by any one of (c, d, e).

(g) $\limsup_{t \to \infty} \dfrac{B(t, \omega)}{\sqrt{2t \log \log t}} = 1$, by (106, Section 5).

(h) $\limsup_{t \downarrow 0} \dfrac{B(t, \omega)}{\sqrt{2t \log \log (1/t)}} = 1$, by (107, Section 5).

Properties (g) and (h) are dual, by (13). According to (Lévy, 1954a, pp. 168–172):

(3) $$\limsup_{h \downarrow 0} L(h, \omega) = 1$$

where

$$L(h, \omega) = \frac{D(h, \omega)}{\sqrt{2h \log (1/h)}}$$

and

$$D(h, \omega) = \max \{|B(s, \omega) - B(t, \omega)| : 0 \leqq s \leqq t \leqq 1 \text{ and } |s - t| < h\}.$$

You should compare (2), (3), and (h). I will not prove (3), which gives an exact modulus of continuity for Brownian motion paths. Most of the results in Sections 2–4 are in (Lévy, 1954, 1965). Other references are (Doob, 1953) and (Itô and McKean, 1965).

In Section 5, I will discuss Strassen's (1964) law of the iterated logarithm for Brownian motion. To state this result, let $C[0, 1]$ be the set of continuous, real-valued functions f on $[0, 1]$. Let

$$\|f\| = \max \{|f(t)| : 0 \leqq t \leqq 1\}.$$

Define the distance between f and g as $\|f - g\|$. Then $C[0, 1]$ is a complete, separable metric space. Let K be the set of absolutely continuous f on $[0, 1]$, with $f(0) = 0$ and

$$\int_0^1 f'(t)^2 \, dt \leqq 1.$$

This K is compact. Call a subset F of $C[0, 1]$ *relatively compact* iff any sequence in F has a convergent subsequence, possibly to some point outside F. This is the same as requiring the closure of F to be compact.

Suppose F is indexed: $F = \{f_n\}$. The *limit set* of F is the set of $g \in C[0, 1]$ such that: for any positive ε, there are arbitrarily large indices n for which f_n is within ε of g. The limit set L of $\{f_n\}$ is closed but may be smaller than the closure of $\{f_n\}$. Indeed, $f_n \in L$ iff either f_n is a non-isolated point of F, or there are arbitrarily large indices m with $f_m = f_n$. The limit set consequently depends on the indexing.

Let

$$Z_n(t, \omega) = \frac{B(nt, \omega)}{\sqrt{2n \log \log n}} \quad \text{for } 0 \leqq t \leqq 1.$$

Thus, $Z_n(\cdot, \omega) \in C[0, 1]$ for $n \geqq 3$. Strassen's iterated logarithm (81) claims:

For almost all ω, the indexed subset

$$\{Z_n(\cdot, \omega): n = 3, 4, \ldots\}$$

of $C[0, 1]$ is relatively compact, with limit set K.

You can get the flavor of this assertion by looking at examples (136 and 138–139).

Here is a fast sketch of the proof.

Fix a positive ε. For almost all ω, for all large n,

$$Z_n(\cdot, \omega) \text{ is within } \varepsilon \text{ of } K.$$

Let ε decrease to 0 through a sequence, discarding the null sets as you go. For the remaining ω, you have forced $\{Z_n(\cdot, \omega)\}$ to be relatively compact, with limit set included in K: because K is compact. To estimate the distance from Z_n to K, introduce the intermediate $Z_{n,m}$ where:

$$Z_{n,m}(k/m) = Z_n(k/m) \quad \text{for } k = 0, \ldots, m$$

$$Z_{n,m} \quad \text{is linearly interpolated.}$$

Estimate the distance from Z_n to $Z_{n,m}$, and from $Z_{n,m}$ to K.

Choose the right countable, dense subset C of K. Fix an f in C and a positive ε. For almost all ω,

$$\|Z_n(\cdot, \omega) - f\| < \varepsilon$$

for infinitely many n. Let ε decrease .ɔ 0 through a sequence and vary f over C, discarding the null sets. For the remaining ω, the limit set of $\{Z_n(\cdot, \omega)\}$ includes all of C. Because limit sets are closed, the limit set of $\{Z_n(\cdot, \omega)\}$ includes all of K. This argument is essentially the one in Strassen's paper. Leonard Cottrell helped me to understand it.

The rest of this chapter is about the ways in which Brownian motion can be used to approximate random walks. The treatment depends on Section 3, but not on the reflection principle or Section 4; Section 5 is used only in Section 8. The main weapon is Skorokhod's (1965) representation, which I will discuss in Section 6. Let F be a distribution function on $(-\infty, \infty)$, with mean 0 and variance $\sigma^2 \leqq \infty$:

$$\int_{-\infty}^{\infty} x \; F(dx) = 0 \quad \text{and} \quad \int_{-\infty}^{\infty} x^2 \; F(dx) = \sigma^2.$$

Remember that B is normalized Brownian motion on $(\Omega, \mathscr{F}, \mathscr{P})$. For technical reasons, suppose $(\Omega, \mathscr{F}, \mathscr{P})$ supports a random variable W which is independent of B and is uniformly distributed over $[0, 1]$. Then (117), there exist times τ_n with properties (a–c).

(a) $0 = \tau_0 \leqq \tau_1 \leqq \tau_2 \leqq \cdots < \infty$.

(b) The differences $B(\tau_{n+1}) - B(\tau_n)$ are independent and have common distribution function F, for $n = 0, 1, \ldots$.

(c) The differences $\tau_{n+1} - \tau_n$ are independent and identically distributed, with mean σ^2.

Call this the *Skorokhod representation* for F. In (c), I really mean mean:

$$E(\tau_{n+1} - \tau_n) = \sigma^2.$$

Why is the Skorokhod representation interesting? First,

$$\{B(\tau_n): n = 0, 1, \ldots\}$$

is a realization of the random walk whose steps have distribution F. Second, if $\sigma^2 < \infty$ then $\tau_n/n\sigma^2 \to 1$ almost surely by the strong law. So, the Skorokhod version of the random walk has trajectory close to that of Brownian motion, with time rescaled by σ^2. In Sections 7 and 8, this leads to proofs of the invariance principles.

To get the main idea of Skorokhod's construction, suppose F concentrates on the two points u and v, with $u < 0 < v$. Define $\tau_0 = 0$, and let τ_{n+1} be the least $t > \tau_n$ with

$$B(t) - B(\tau_n) = u \quad \text{or} \quad v.$$

Use martingales (114) to show that $B(\tau_1)$ has distribution function F, while τ_1 has mean $|u|v$. Then use strong Markov to finish this case off. The construction for $u = -1$ and $v = 1$ is illustrated by Figure 2.

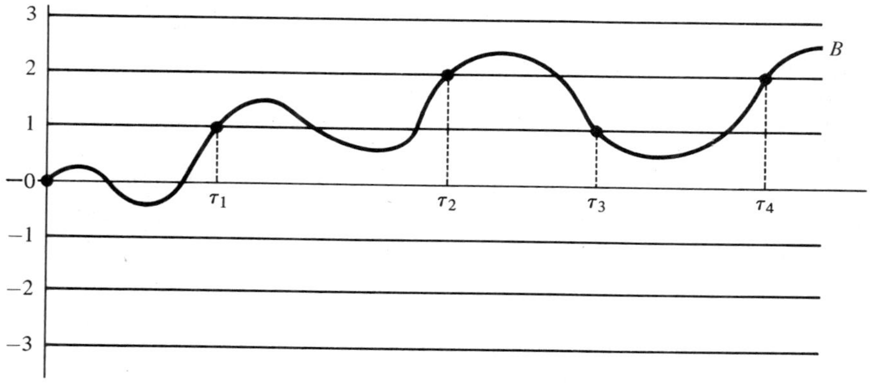

The Skorokhod representation.
Figure 2.

General F can be represented as the average of two-point, mean 0 distributions (108). Now the general case can be done by randomizing u and v at each stage n, independently in n, so the average of the mean 0 distribution sitting on random u and v is F. I learned this argument from Volker Strassen. Section 6 closes with (118–119), which were used in Chapter 3 of *MC*.

Section 7 is concerned with Donsker's (1951) invariance principle. The treatment depends on Sections 3 and 6, but not on the reflection principle or Sections 4, 5. Suppose $X_1, X_2, \ldots$ are independent and identically distributed, having mean 0 and variance 1. Let $S_0 = 0$, and let

$$S_n = X_1 + \cdots + X_n$$

for $n \geqq 1$. The central limit theorem asserts that S_n is asymptotically normal: the distribution of $S_n/n^{\frac{1}{2}}$ converges to $N(0, 1)$. In what sense? Let π be the $N(0, 1)$ distribution, so π is a probability on $(-\infty, \infty)$. Then

$$E\{\varphi(S_n/n^{\frac{1}{2}})\} \to \int_{-\infty}^{\infty} \varphi(z)\, \pi(dz)$$

for any bounded, continuous function φ on $(-\infty, \infty)$. Donsker asserts that the whole trajectory $(S_0, \ldots, S_n)$ is asymptotically normal: its distribution after rescaling converges to that of normalized Brownian motion B. Now $(S_0, \ldots, S_n)$ is a discrete process, and B is continuous. To fix this up, join the points

$$(0, S_0), (1, S_1), \ldots, (n, S_n)$$

by straight line segments, to make a continuous curve C_n. To control the spread of S_n, the central limit theorem divides the space scale by $n^{\frac{1}{2}}$. This suggests looking at $C_n/n^{\frac{1}{2}}$. But

$$C_n(n)/n^{\frac{1}{2}}$$

is almost distributed like $B(1)$. This suggests dividing the time scale of C_n by n, and confining the time parameter of B to $[0, 1]$.

For the first operation, introduce

$$S_{(n)}(t) = C_n(nt)/n^{\frac{1}{2}} \quad \text{for } 0 \leqq t \leqq 1.$$

More generally, suppose $\{s_n : n = 0, 1, \ldots\}$ is a sequence of real numbers. Then $s_{(n)} \in C[0, 1]$ is defined by the requirements:

$$s_{(n)}(i/n) = s_i/n^{\frac{1}{2}} \qquad \text{for } i = 0, \ldots, n;$$

$$s_{(n)} \quad \text{is linear on} \left[\frac{i-1}{n}, \frac{i}{n}\right] \qquad \text{for } i = 1, \ldots, n.$$

And

$$S_{(n)}(\cdot, \omega) = S_{(n)}(\omega).$$

Of course, $S_{(n)}$ is a random element of $C[0, 1]$. The definition of $S_{(n)}$ is

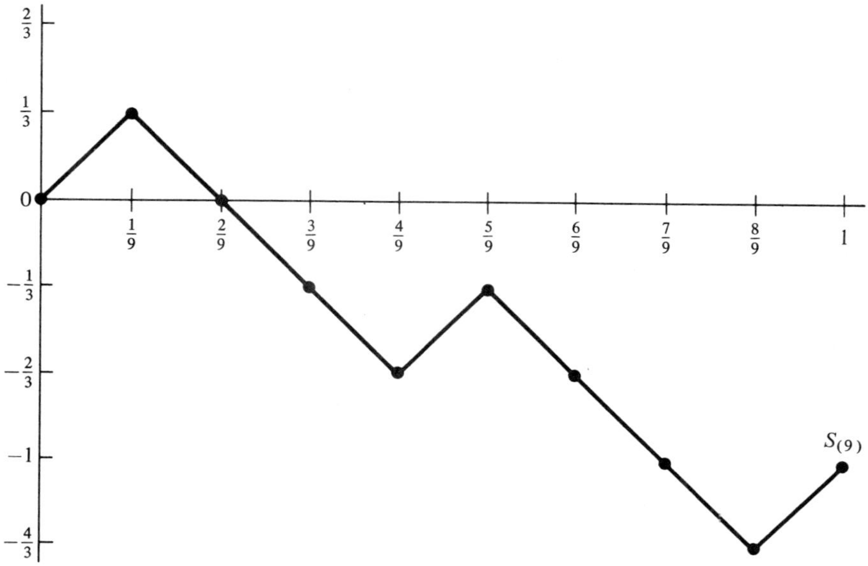

Figure 3.

illustrated by Figure 3, in the case $X_m = \pm 1$ with probability $\frac{1}{2}$ each, and $n = 9$. The X-sequence was generated by tossing a nickel, Jefferson counting as 1 and Monticello as -1. The values obtained for S_0 through S_9 were

$$0, 1, 0, -1, -2, -1, -2, -3, -4, -3.$$

And $S_{(9)}$ is graphed in Figure 3.

For the second operation, introduce the distribution π_1 of

$$\{B(t): 0 \leqq t \leqq 1\},$$

where B is still normalized Brownian motion. Of course, π_1 is a probability on $C[0, 1]$. And Donsker (120) asserts

$$E\{\varphi(S_{(n)})\} \to \int_{C[0,1]} \varphi(f)\, \pi_1(df)$$

for any bounded, continuous function φ on $C[0, 1]$. What does continuous mean? I already gave $C[0, 1]$ the metric

$$\text{distance}\,(f, g) = \|f - g\|,$$

where

$$\|h\| = \max\,\{|h(t)| : 0 \leqq t \leqq 1\}.$$

And I mean continuous relative to this metric. To get the flavor, look at examples (121–124).

Technical aside. The central limit theorem also asserts that

$$E\{\varphi(S_{(n)})\} \to \int \varphi \, d\pi_1,$$

for a narrower class of bounded functionals φ. This class contains all the φ which depend continuously on the values of f at a finite set of times; this set is allowed to depend on φ. And the class is closed under passages to the limit which are uniform on the spheres of $C[0, 1]$: estimate (5) keeps $\|S_{(n)}\|$ under control. Jack Feldman proved to me that the closure of the finite dimensional φ, in the sense I just described, coincides with the class of φ which are uniformly continuous on the spheres of $C[0, 1]$, provided $C[0, 1]$ is equipped with the topology of pointwise convergence.

Here is an outline of the proof for Donsker. I only have to do it when φ is uniformly continuous: because general φ can be approximated above and below by uniformly continuous φ. This part of the argument is isolated in (125–127). Let F be the distribution function of the steps X_m in the random walk, and let τ_n be the times in the Skorokhod representation for F. Because the variance is 1,

$$\tau_n/n \to 1 \quad \text{almost surely.}$$

Let

$$\hat{S}_n = B(\tau_n).$$

Define $\hat{S}_{(n)}$ by the usual requirements:

$$\hat{S}_{(n)}(j/n) = \hat{S}_j/n^{\frac{1}{2}}$$

$$\hat{S}_{(n)} \quad \text{is linearly interpolated.}$$

Let

$$B_n(t) = B(nt)/n^{\frac{1}{2}} \quad \text{for } 0 \leqq t \leqq 1,$$

so B_n is a random element of $C[0, 1]$. By estimate (19),

$$\|\hat{S}_{(n)} - B_n\| \to 0$$

in probability. Because φ is bounded and uniformly continuous, (129) forces

$$E\{\varphi(\hat{S}_{(n)})\} - E\{\varphi(B_n)\} \to 0.$$

Because $\{\hat{S}_n\}$ is distributed like $\{S_n\}$,

$$E\{\varphi(\hat{S}_{(n)})\} = E\{\varphi(S_{(n)})\}.$$

By scaling (11),

$$E\{\varphi(B_n)\} = E\{\varphi(B_1)\} = \int \varphi \, d\pi_1.$$

In Section 8, I will discuss Strassen's (1964) invariance principle. The treatment depends on Sections 3, 5, and 6, but not on the reflection principle or Section 4. Suppose $X_1, X_2, \ldots$ are independent and identically distributed, with mean 0 and variance 1. Let $S_0 = 0$ and

$$S_n = X_1 + \cdots + X_n.$$

Define $S_{(n)}$ as for Donsker:

$$S_{(n)}(j/n) = S_j/n^{\frac{1}{2}} \quad \text{for } j = 0, \ldots, n$$

$$S_{(n)} \quad \text{is linearly interpolated.}$$

Let

$$T_n = S_{(n)}/(2 \log \log n)^{\frac{1}{2}}.$$

Thus, T_n is a random element of $C[0, 1]$, and is another rescaling of the random walk trajectory. In particular,

$$T_n(1) = S_n/(2n \log \log n)^{\frac{1}{2}},$$

the usual object of study for the law of the iterated logarithm. Define K as in Section 5: the set of absolutely continuous functions f on $[0, 1]$, with $f(0) = 0$ and

$$\int_0^1 f'(t)^2 \, dt \leqq 1.$$

Strassen's invariance principle (133) states:

> For almost all sample points, the indexed subset $\{T_n : n \geqq 3\}$ of $C[0, 1]$ is relatively compact, with limit set K.

To get the hang of this, look at examples (136 and 138–139).

Here is a sketch of the argument. Let F be the distribution function of the steps X_m. Let $\{\tau_n\}$ be the times in the Skorokhod representation for F. As before, let

$$\hat{S}_n = B(\tau_n).$$

Define $\hat{S}_{(n)}$ by the usual requirements:

$$\hat{S}_{(n)}(j/n) = S_j/n^{\frac{1}{2}} \quad \text{for } j = 0, \ldots, n$$

$$\hat{S}_{(n)} \quad \text{is linearly interpolated.}$$

Let

$$\hat{T}_n = \hat{S}_{(n)}/(2 \log \log n)^{\frac{1}{2}}.$$

Remember

$$Z_n(t) = B(nt)/(2n \log \log n)^{\frac{1}{2}} \quad \text{for } 0 \leqq t \leqq 1.$$

Using estimate (20),

$$\|\hat{T}_n - Z_n\| \to 0 \quad \text{almost surely.}$$

Now Strassen's iterated logarithm for Brownian motion (Section 5) implies his invariance principle, with $\hat{T}_n$ in place of T_n. But $\{\hat{S}_n\}$ is distributed like $\{S_n\}$, so $\{\hat{T}_n\}$ is distributed like $\{T_n\}$. These ideas are developed further in (Strassen, 1966, 1967).

The results of Sections 7 and 8 extend in a natural way to martingales with bounded increments. One such extension, a variant (143) of Lindeberg's theorem, is in Section 9. Section 10 is an aside on random sums. As usual, let $X_1, X_2, \ldots$ be independent and identically distributed, with mean 0 and variance 1. Let $S(0) = 0$ and

$$S(n) = X_1 + \cdots + X_n.$$

Let $\tau_1, \tau_2, \ldots$ be a sequence of nonnegative, integer-valued random variables. Do not make any assumption about the dependence of the τ's and the X's: do not even assume the τ's are stopping times. Instead, suppose that τ_n/n converges in probability to a positive, finite, limiting variable. Then (Wittenberg, 1964):

$$\text{the distribution of } S(\tau_n)/\tau_n^{\frac{1}{2}} \text{ converges to } N(0, 1).$$

This is proved as (150). Section 11 is on Wald's identity for Brownian motion. On a first reading of the book, you might skip Sections 9–11; the argument in Section 9 is rather condensed.

2. FOUNDATIONS

The main result (6) of this section, due to Wiener (1923), is that normalized Brownian motion exists. There are two preliminary results (4, 5). The first is from (Feller, 1968, p. 175). The idea of the second is due to Lévy; I learned its elegant formulation from Leo Breiman.

To state (4), let Y be $N(0, 1)$ on the triple $(\Omega, \mathscr{F}, \mathscr{P})$.

(4) Lemma.

(a) $$\frac{1}{y} - \frac{1}{y^3} \leqq (2\pi)^{\frac{1}{2}} \cdot e^{\frac{1}{2}y^2} \cdot \mathscr{P}\{Y \geqq y\} \leqq \frac{1}{y} \quad \text{for } y > 0.$$

(b) *Let* $-\infty < a < b < \infty$ *and let* $m = \max\{a^2, b^2\}$. *Then*

$$(b - a)e^{-m} \leqq (2\pi)^{\frac{1}{2}} \cdot \mathscr{P}\{a \leqq Y \leqq b\} \leqq (b - a).$$

PROOF. *Claim (a).* I say

$$e^{-\frac{1}{2}y^2}\left(\frac{1}{y} - \frac{1}{y^3}\right) = \int_y^\infty e^{-\frac{1}{2}x^2}\left(1 - \frac{3}{x^4}\right) dx.$$

Indeed, both sides tend to 0 as y tends to ∞, and both sides have the same derivative for all $y > 0$. The display shows

$$e^{-\frac{1}{2}y^2}\left(\frac{1}{y} - \frac{1}{y^3}\right) \leqq \int_y^\infty e^{-\frac{1}{2}x^2}\,dx = (2\pi)^{\frac{1}{2}}\,\mathscr{P}\{Y \geqq y\}.$$

For the other inequality, use

$$e^{-\frac{1}{2}y^2}\frac{1}{y} = \int_y^\infty e^{-\frac{1}{2}x^2}\left(1 + \frac{1}{x^2}\right)dx.$$

Claim (*b*). The $N(0, 1)$ density on $[a, b]$ is between $(2\pi)^{-\frac{1}{2}}e^{-m}$ and $(2\pi)^{-\frac{1}{2}}$. ★

To state (5), let $Y_1, \ldots, Y_n$ be independent random variables on $(\Omega, \mathscr{F}, \mathscr{P})$. Let

$$S_j = Y_1 + \cdots + Y_j \quad\text{and}\quad T_j = Y_{j+1} + \cdots + Y_n,$$

so $S_0 = T_n = 0$. Let $a \leqq 0 \leqq b$. Let

$$B_j = \{T_j \geqq a\} \quad\text{and}\quad \gamma = \min_{0 \leqq j \leqq n} \mathscr{P}\{B_j\}.$$

(5) Lemma. $\gamma \cdot \mathscr{P}\{\max_{0 \leqq j \leqq n} S_j \geqq b\} \leqq \mathscr{P}\{S_n \geqq a + b\}$.

PROOF. Let $A_0 = \{S_0 \geqq b\}$. For $j \geqq 1$, let

$$A_j = \{S_0 < b, \ldots, S_{j-1} < b, S_j \geqq b\}.$$

The A_j are disjoint; and

$$\bigcup_{j=0}^n A_j = \{\max_{0 \leqq j \leqq n} S_j \geqq b\};$$

while A_j and B_j are independent; finally,

$$\bigcup_{j=0}^n A_j B_j \subset \{S_n \geqq a + b\}.$$

Therefore,

$$\begin{aligned}\gamma \cdot \mathscr{P}\{\max_{0 \leqq j \leqq n} S_j \geqq b\} &= \gamma \textstyle\sum_{j=0}^n \mathscr{P}\{A_j\}\\ &\leqq \textstyle\sum_{j=0}^n \mathscr{P}\{A_j B_j\}\\ &\leqq \mathscr{P}\{S_n \geqq a + b\}.\end{aligned}$$

★

(6) Theorem. *Normalized Brownian motion exists.*

PROOF. Let Ω be the set of all functions ω from the binary rationals R in $[0, \infty)$ to the real line, such that $\omega(0) = 0$. Let $X(r, \omega) = \omega(r)$ for $\omega \in \Omega$ and $r \in R$. Let $\mathscr{F}$ be the smallest σ-field of subsets of Ω such that $X(r)$ is measurable for all r. There is a probability $\mathscr{P}$ on $\mathscr{F}$, for example by Kolmogorov consistency (3.53), such that (1c) holds for binary rational $t_1, \ldots, t_n$. For

$K < \infty$, let $C(K)$ be the set of $\omega \in \Omega$ such that $X(\cdot, \omega)$ is uniformly continuous on $[0, K] \cap R$. You should verify $C(K) \in \mathscr{F}$. The main problem in (6) is proving

(7) $$\mathscr{P}\{C(K)\} = 1.$$

Without real loss, set $K = 1$. The informal idea is to study the 2^k-grid

$$0/2^k, 1/2^k, 2/2^k, \ldots, (2^k - 1)/2^k, 2^k/2^k$$

of $[0, 1]$, and to argue that for large enough random k: for all $i = 0, \ldots, 2^k - 1$, the values of the process on $[i/2^k, (i + 1)/2^k]$ do not differ much from the value at the left endpoint $i/2^k$. To carry out this idea, study the finer 2^n-grid, and argue from (5) that at the fine grid points, the process is close to its value at the nearest 2^k-grid point on the left.

Formally, let $n \geqq 1$ and $b_k > 0$. Rationalize the next computation this way. In line 1, use (5) with $Y_j = X(j/2^n) - X((j - 1)2^n)$; put $a = 0$, so $\gamma = \frac{1}{2}$; put $b = b_k$. In line 2, use the definitions. In line 3, use (4a).

$$\begin{aligned}\mathscr{P}\{b_k \leqq \max\,[X(j/2^n)\colon j = 0, \ldots, 2^{n-k}]\} &\leqq 2\mathscr{P}\{X(1/2^k) \geqq b_k\}\\ &= 2\mathscr{P}\{X(1) \geqq b_k\sqrt{2^k}\}\\ &\leqq \beta_k,\end{aligned}$$

where β_k is

$$\sqrt{\frac{2}{\pi}} \cdot \frac{e^{-2^{k-1}b_k^2}}{b_k\sqrt{2^k}}.$$

By symmetry,

(8) $$\mathscr{P}\{b_k \leqq \max\,[|X(j/2^n)| : j = 0, \ldots, 2^{n-k}]\} \leqq 2\beta_k.$$

Let $A(i, k, n)$ be the event that

$$b_k \leqq \max\,\{|X(i2^{-k} + j2^{-n}) - X(i2^{-k})| : j = 0, \ldots, 2^{n-k}\}.$$

By (8) and (1c),

$$\mathscr{P}\{A(i, k, n)\} \leqq 2\beta_k.$$

Let

$$A(k, n) = \bigcup\,\{A(i, k, n) : i = 0, \ldots, 2^k - 1\},$$

so

$$\mathscr{P}\{A(k, n)\} \leqq 2^{k+1}\beta_k.$$

As n increases, $A(k, n)$ increases, say to $A(k)$; so

$$\mathscr{P}\{A(k)\} \leqq 2^{k+1}\beta_k.$$

Now

$$C(1) \supset \liminf \Omega \setminus A(k),$$

provided $b_k \to 0$. If b_k is chosen properly, say $b_k = (k2^{-k+1})^{\frac{1}{2}}$, then $\Sigma_k 2^{k+1}\beta_k < \infty$; Borel Cantelli now proves (7).

Let $C = \bigcap_{K=1}^{\infty} C(K)$, so $\mathscr{P}\{C\} = 1$. For $\omega \in C$, define $B(\cdot, \omega)$ as the unique continuous function on $[0, \infty)$ which extends $X(\cdot, \omega)$ on R. For $\omega \notin C$, define $B(\cdot, \omega) \equiv 0$, say. Property (1c) follows by continuity. ★

To state the next result, let I be a subinterval of $[0, \infty)$; in application, $I = [0, \infty)$, $[0, 1]$, $(0, \infty)$. Let $C(I)$ be the set of continuous functions f on I. For $f \in C(I)$ and $t \in I$, let $\xi(t, f) = f(t)$. Give $C(I)$ the smallest σ-field of subsets for which all $\xi(t)$ are measurable.

(9) Lemma. *Let P and Q be two probabilities on $C(I)$. Suppose that for all n and all $t_1, \ldots, t_n$ in I with $t_1 < \cdots < t_n$, the P-distribution of $\xi(t_1), \ldots, \xi(t_n)$ coincides with the Q-distribution of $\xi(t_1), \ldots, \xi(t_n)$. Then $P = Q$.*

PROOF. Use (3.16). ★

The following are measurable maps from $(\Omega, \mathscr{F})$ to $C[0, \infty)$, where $\sigma > 0$ and $s \geqq 0$ and $-\infty < x < \infty$:

(a) B, which sends $\omega \in \Omega$ to the function in $C[0, \infty)$ whose value at t is $B(t, \omega)$;

(b) $-B$, which sends $\omega \in \Omega$ to the function in $C[0, \infty)$ whose value at t is $-B(t, \omega)$;

(c) $\sigma^{-1}B(\sigma^2 \cdot)$, which sends $\omega \in \Omega$ to the function in $C[0, \infty)$ whose value at t is $\sigma^{-1}B(\sigma^2 t, \omega)$;

(d) $x + B$, which sends $\omega \in \Omega$ to the function in $C[0, \infty)$ whose value at t is $x + B(t, \omega)$;

(e) $B(s + \cdot)$, which sends $\omega \in \Omega$ to the function in $C[0, \infty)$ whose value at $t \in [0, \infty)$ is $B(s + t, \omega)$;

(f) $B(s + \cdot) - B(s)$, which sends $\omega \in \Omega$ to the function in $C[0, \infty)$ whose value at $t \in [0, \infty)$ is $B(s + t, \omega) - B(s, \omega)$.

Let $s > 0$. Then $\{B(u): 0 \leqq u \leqq s\}$ is a measurable map from $(\Omega, \mathscr{F})$ to $C[0, s]$; its value at $\omega \in \Omega$ is the function $B(\cdot, \omega)$ retracted to $[0, s]$. Likewise, $\{B(s - u) - B(s): 0 \leqq u \leqq s\}$ is a measurable map from $(\Omega, \mathscr{F})$ to $C[0, s]$, which sends $\omega \in \Omega$ to the function in $C[0, s]$ whose value at $u \in [0, s]$ is

$$B(s - u, \omega) - B(s, \omega).$$

(10) Definition. *If M is a measurable map from $(\Omega, \mathscr{F})$ to $C(I)$, the $\mathscr{P}$-distribution of M is the probability $\mathscr{P}M^{-1}$ on $C(I)$. Let P_x be the $\mathscr{P}$-distribution of $x + B$.*

The *symmetry* (11a) and *scaling* (11b) properties of B will be useful. So will (11c) and (11d). To state (11c) and (12), let $\mathscr{F}(s)$ be the σ-field spanned by $B(u)$ for $0 \leqq u \leqq s$.

(11) Lemma. **(a)** *The $\mathscr{P}$-distribution of $-B$ is P_0.*

(b) *The $\mathscr{P}$-distribution of $\sigma^{-1}B(\sigma^2 \cdot)$ is P_0.*

(c) *$B(s + \cdot) - B(s)$ is $\mathscr{P}$-independent of $\mathscr{F}(s)$, and has $\mathscr{P}$-distribution P_0.*

(d) *The $\mathscr{P}$-distribution of*

$$\{B(u): 0 \leqq u \leqq s\}$$

coincides with the $\mathscr{P}$-distribution of

$$\{B(s - u) - B(s): 0 \leqq u \leqq s\}.$$

PROOF. Use (9) and definition (1). ★

NOTE. The second process in (d) is obtained by taking B on $[0, s]$, reversing time, and renormalizing the reversed process to start at 0.

Lemma (11c) can be restated as the *Markov property* (12). Regular conditional distributions are discussed in Section 3.10.

(12) Theorem. *A regular conditional $\mathscr{P}$-distribution for $B(s + \cdot)$ given $\mathscr{F}(s)$ is $P_{B(s)}$: namely, the function*

$$(\omega, A) \to P_{B(s,\omega)}(A)$$

of $\omega \in \Omega$ and measurable subsets A of $C[0, \infty)$.

PROOF. Let $G \in \mathscr{F}(s)$ and let A be a measurable subset of $C[0, \infty)$. I have to show

$$\mathscr{P}\{G \text{ and } B(s + \cdot) \in A\} = \int_G P_{B(s,\omega)}(A)\, \mathscr{P}(d\omega).$$

For real x, and $f \in C[0, \infty)$, the function $x + f \in C[0, \infty)$ takes the value $x + f(t)$ at $t \in [0, \infty)$. The mapping

$$(x, f) \to (x + f)$$

is product measurable; and the P_0-distribution of

$$f \to x + f$$

is P_x. Now consider the product space

$$\Omega \times C[0, \infty),$$

where Ω is given the σ-field $\mathscr{F}(s)$ and $C[0, \infty)$ is given the usual σ-field. Let G^* be the set of pairs (ω, f) such that

$$\omega \in G \quad \text{and} \quad B(s, \omega) + f \in A.$$

Then G^* is product measurable, because $B(s)$ and G are $\mathscr{F}(s)$-measurable. The ω-section $G^*(\omega)$ of G^* is empty for $\omega \notin G$, and is

$$\{f : f \in C[0, \infty) \text{ and } B(s, \omega) + f \in A\}$$

for $\omega \in G$. Use (11c) and Fubini (3.21): put $(\Omega, \mathscr{F}(s))$ for $(\Omega_1, \mathscr{F}_1)$, and $X_1(\omega) = \omega$; put $C[0, \infty)$ for $(\Omega_2, \mathscr{F}_2)$, and $X_2(\omega) = B(s + \cdot, \omega) - B(s, \omega)$.

$$\begin{aligned} \mathscr{P}\{G \text{ and } B(s + \cdot) \in A\} &= \mathscr{P}\{[X_1(\omega), B(s + \cdot, \omega) - B(s, \omega)] \in G^*]\} \\ &= \int P_0\{G^*(\omega)\}\, \mathscr{P}(d\omega) \\ &= \int_G P_0\{f : B(s, \omega) + f \in A\}\, \mathscr{P}(d\omega) \\ &= \int_G P_{B(s,\omega)}(A)\, \mathscr{P}(d\omega). \end{aligned}$$

★

The next result (13) shows that 0 and ∞ are dual. To state it, define Y and Z as follows, for $0 < t < \infty$ and $\omega \in \Omega$:

$$Y(t, \omega) = t^{-\frac{1}{2}} B(t, \omega) \quad \text{and} \quad Z(t, \omega) = Y(t^{-1}, \omega) = t^{\frac{1}{2}} B(t^{-1}, \omega).$$

(13) Theorem. *Y and Z have the same distribution.*

The proof of (13) requires a small detour (14–15). If M is a matrix, M^* is its *transpose*: $M^*(i, j) = M(j, i)$. Recall that N is orthonormal iff N is square and $NN^* = I$, the identity matrix. Then $N^*N = I$ also.

(14) Lemma. *Let L and M be $n \times n$ matrices. There is an orthonormal $n \times n$ matrix N with $NL = M$ iff $L^*L = M^*M$.*

PROOF. The "only if" part is trivial: $(NL)^*NL = L^*N^*NL = L^*L$. For "if", let $L_1, \ldots, L_n$ be the columns of L, and $M_1, \ldots, M_n$ the columns of M. Let $\langle u, v \rangle$ be the inner product of n-vectors u and v, namely

$$\langle u, v \rangle = \Sigma_{i=1}^n u(i)v(i).$$

Clearly, the (i, j) entry in L^*L is $\langle L_i, L_j \rangle$. Define a formal mapping N_0 as follows:

$$N_0(\Sigma_{i=1}^n c_i L_i) = \Sigma_{i=1}^n c_i M_i.$$

Let $u = \Sigma_{i=1}^n c_i L_i$ and $v = \Sigma_{i=1}^n d_i L_i$. Then $\langle u, v \rangle = \langle N_0 u, N_0 v \rangle$ because $L^*L = M^*M$. If $u = v$, then $N_0 u = N_0 v$: by linearity, you only have to check that $u = 0$ entails $N_0 u = 0$. But $u = 0$ implies $\langle u, u \rangle = 0$ implies $\langle N_0 u, N_0 u \rangle = 0$ implies $N_0 u = 0$. Thus N_0 is well defined, and is an isometry of the column space of L onto the column space of M. Consequently, it can be extended to an isometry N_1 of the space of column n-vectors. Plainly, $N_1(v) = Nv$, where N is an orthonormal matrix. Moreover, $M = NL$. ★

Let $V = (V_1, \ldots, V_n)$ be a row n-vector of random variables. Then V is *centered Gaussian* iff there exists a row n-vector U of independent $N(0, 1)$ random variables and an $n \times n$ matrix L such that V is distributed like UL. In particular, $E(U) = 0$. Define cov V as the expectation of V^*V; that is, $(\text{cov } V)(i, j) = E(V_i V_j)$. Check $\text{cov}(UL) = L^*L$, because cov U is the identity matrix.

(15) Lemma. *Suppose V is centered Gaussian. Then the distribution of V is determined by* cov V.

PROOF. Suppose V is distributed like UL, where U is a row n-vector of independent $N(0, 1)$ variables. Suppose furthermore that another row n-vector T is distributed like SM, where S is also a row n-vector of independent $N(0, 1)$ variables. Suppose finally that cov V = cov T. The problem is to show that V and T have the same distribution. Now cov $V = L^*L$, and cov $T = M^*M$. As (14) implies, $M = NL$, where N is orthonormal. Therefore, T is distributed like $(SN)L$. But SN is distributed like U. Indeed, the joint density of S at the row n-vector x is

$$(2\pi)^{-(n/2)} \cdot e^{-\frac{1}{2}\|x\|^2},$$

where $\|x\|^2 = \langle x, x \rangle$ is invariant under N. ★

PROOF OF (13). Check that $E[B(t)B(s)] = \min(t, s)$. Then compute $E[Y(t)Y(s)]$ and $E[Z(t)Z(s)]$, checking that the two expectations are equal. Let $0 < t_1 < t_2 < \cdots < t_n$. I say that

$$[Y(t_1), Y(t_2), \ldots, Y(t_n)]$$

is centered Gaussian, as is

$$[Z(t_1), Z(t_2), \ldots, Z(t_n)].$$

For instance, the Z-vector is $(U_1, \ldots, U_n)$ times some matrix, where

$$U_m = \left(\frac{1}{t_m} - \frac{1}{t_{m+1}}\right)^{-\frac{1}{2}} \cdot \left[B\left(\frac{1}{t_m}\right) - B\left(\frac{1}{t_{m+1}}\right)\right],$$

and $1/t_{n+1}$ is temporarily 0. The Y-vector and the Z-vector then have the same distribution, by (15). Finally, use (9). ★

Three of the estimates in the proof of (6) are worth preserving. For precise results, see (26) and (34).

(16) Lemma. *Let* $b \geqq 0$. *Then*

(a) $\mathscr{P}\{\max_{0 \leqq s \leqq t} B(s) \geqq b\} \leqq 2\,\mathscr{P}\{B(t) \geqq b\}$

(b) $\mathscr{P}\{\min_{0 \leqq s \leqq t} B(s) \leqq -b\} \leqq 2\,\mathscr{P}\{B(t) \leqq -b\}$

(c) $\mathscr{P}\{\max_{0 \leqq s \leqq t} |B(s)| \geqq b\} \leqq 2\,\mathscr{P}\{|B(t)| \geqq b\} = 4\,\mathscr{P}\{B(1) \geqq bt^{-\frac{1}{2}}\}$.

NOTE. $\max_{0 \leqq s \leqq t} B(s) = \sup\{B(r) : 0 \leqq r \leqq t$ and r is a binary rational$\}$. So everything is measurable.

PROOF. I will argue the first estimate; the second is symmetric by (11a); and adding the first two gives the third. Without real loss (11b), put $t = 1$. If $b = 0$, the result is trivial, so assume $b > 0$. Let $0 < b' < b$. Let A_n be the event

$$b' \leqq \max\{B(j/2^n) : j = 0, \ldots, 2^n\}.$$

By (5),

$$\mathscr{P}\{A_n\} \leqq 2\,\mathscr{P}\{B(1) \geqq b'\}.$$

Now A_n increases with n, say to A; and

$$\{\max_{0 \leqq s \leqq 1} B(s) \geqq b\} \subset A.$$

So,

$$\mathscr{P}\{\max_{0 \leqq s \leqq 1} B(s) \geqq b\} \leqq 2\,\mathscr{P}\{B(1) \geqq b'\}.$$

Let b' increase to b. ★

(17) Lemma. $\mathscr{P}\{\limsup_{n \to \infty} B(n) = \infty\} = 1$

$\mathscr{P}\{\liminf_{n \to \infty} B(n) = -\infty\} = 1.$

CLUMSY PROOF. Let r be a large, positive integer, like 9. Use (4) and Borel-Cantelli: for almost all ω,

$$B(r^k, \omega) \geqq -3\sqrt{r^k \log k} \quad \text{for all but finitely many } k$$

$$B(r^{k+1}, \omega) - B(r^k, \omega) \geqq \sqrt{2(r-1)r^k \log k} \quad \text{for infinitely many } k.$$

For a good ω,

$$B(r^{k+1}, \omega) \geqq [\sqrt{2(r-1)} - 3]\sqrt{r^k \log k} \quad \text{for infinitely many } k,$$

doing in the lim sup. The lim inf is symmetric by (11a). ★

SLICK PROOF. Fix a positive integer m. Then $\mathscr{P}\{B(n) \geqq m\} \to 1$, so

$$\mathscr{P}\{B(n) \geqq m \quad \text{for infinitely many } n\} \geqq \tfrac{1}{2},$$

by Fatou. Let $m \uparrow \infty$ and use countable additivity:

$$\mathscr{P}\{\limsup B(n) = \infty\}$$

is at least $\frac{1}{2}$. The Hewitt-Savage 0-1 Law (*MC*, 1.122) now forces this probability up to 1. ★

The general situation is described in (*MC*, 3.35).

(18) Important convention. From now on, B is normalized Brownian motion on the triple $(\Omega, \mathscr{F}, \mathscr{P})$. And

$$\limsup_{n\to\infty} B(n,\omega) = +\infty$$

$$\liminf_{n\to\infty} B(n,\omega) = -\infty$$

for all $\omega \in \Omega$.

The next two estimates will be used in Sections 5–8.

(19) Lemma. *Let $\varepsilon > 0$. Let $r > 1$. The $\mathscr{P}$-probability that*

$$\varepsilon\sqrt{n} < \max\{|B(t) - B(s)| : 0 \leqq s \leqq n \text{ and } s \leqq t \leqq rs\}$$

does not depend on n, and converges to 0 as r decreases to 1.

PROOF. Use the scaling property (11b), and the continuity of $B(\cdot, \omega)$. ★

(20) Lemma. *Let $\varepsilon > 0$. There is an $r > 1$ such that*

$$\mathscr{P}\{\limsup A_n\} = 0,$$

where A_n is the event that

$$\max\{|B(t) - B(s)| : 0 \leqq s \leqq n \text{ and } s \leqq t \leqq rs\}$$

exceeds $\varepsilon(n \log\log n)^{1/2}$.

PROOF. I will choose $r > 1$ soon. Let E_n be the event that

$$\max\{|B(t) - B(s)| : 1 \leqq s \leqq n \text{ and } s \leqq t \leqq rs\}$$

exceeds

$$\varepsilon(n \log\log n)^{\frac{1}{2}}$$

I only have to show $\mathscr{P}\{\limsup E_n\} = 0$. Let

$$F_k = \bigcup\{E_n : r^k \leqq n \leqq r^{k+1}\}.$$

By Borel-Cantelli, it is enough to show $\Sigma_k \mathscr{P}\{F_k\} < \infty$. I will argue that

$$\mathscr{P}\{F_k\} = 0(k^{-a}), \quad \text{where} \quad a = \varepsilon^2/[8(r^2 - 1)] > 1$$

for small $r > 1$. Let

$$\alpha_k = \varepsilon[r^k(\log k + \log\log r)]^{\frac{1}{2}}.$$

Let $F_{k,j}$ be the event that

$$\tfrac{1}{2}\alpha_k \leqq \max\{|B(t) - B(r^j)| : r^j \leqq t \leqq r^{j+2}\}.$$

On F_k: there are s and t with $s \leqq t \leqq rs$ and $1 \leqq s \leqq r^{k+1}$ and

$$|B(t) - B(s)| \geqq \alpha_k;$$

if $r^j \leqq s \leqq r^{j+1}$, then $r^j \leqq s, t \leqq r^{j+2}$, and

$$\text{either} \quad |B(t) - B(r^j)| \geqq \tfrac{1}{2}\alpha_k \quad \text{or} \quad |B(s) - B(r^j)| \geqq \tfrac{1}{2}\alpha_k.$$

Consequently,

$$F_k \subset \cup_{j=0}^k F_{k,j}, \quad \text{and} \quad \mathscr{P}\{F_k\} \leqq \Sigma_{j=0}^k \mathscr{P}\{F_{k,j}\}.$$

Let

$$A(t) = \max\{|B(s)| : 0 \leqq s \leqq t\}$$

and

$$\beta_{k,j} = \tfrac{1}{2}\alpha_k/[r^j(r^2 - 1)]^{\frac{1}{2}}.$$

Then

$$\begin{aligned} \mathscr{P}\{F_{k,j}\} &= \mathscr{P}\{A(r^{j+2} - r^j) \geqq \tfrac{1}{2}\alpha_k\} && (11c)\\ &\leqq 4\,\mathscr{P}\{B(1) \geqq \beta_{k,j}\} && (16c)\\ &\leqq \frac{4}{\sqrt{2\pi}} \cdot \frac{1}{\beta_{k,j}} \cdot e^{-\frac{1}{2}\beta_{k,j}^2} && (4a). \end{aligned}$$

Now

$$\begin{aligned} \Sigma_{j=0}^k 1/\beta_{k,j} &= \frac{2}{\alpha_k}(r^2 - 1)^{\frac{1}{2}} \cdot \Sigma_{j=0}^k r^{\frac{1}{2}j} \\ &= \frac{2}{\alpha_k}(r^2 - 1)^{\frac{1}{2}} \cdot \frac{r^{\frac{1}{2}(k+1)} - 1}{r^{\frac{1}{2}} - 1} \\ &= 0(1) \end{aligned}$$

as $k \to \infty$, for fixed r and ε. Clearly, $\beta_{k,j} \geqq \beta_{k,k}$ for $j = 0, \ldots, k$. So

$$e^{-\frac{1}{2}\beta_{k,j}^2} \leqq e^{-\frac{1}{2}\beta_{k,k}^2} = 0(k^{-a}),$$

as $k \to \infty$; the constant depends on r and ε; while

$$a = \varepsilon^2/8(r^2 - 1).$$

Now ε is fixed, so $a > 1$ for sufficiently small $r > 1$, forcing

$$\begin{aligned} \mathscr{P}\{F_k\} &\leqq \Sigma_{j=0}^k \mathscr{P}\{F_{k,j}\} \\ &= \text{const.}\,[\Sigma_{j=0}^k 1/\beta_{k,j}]\, e^{-\frac{1}{2}\beta_{k,k}^2} \\ &\leqq \text{const.}\, k^{-a} \qquad \text{as } k \to \infty. \end{aligned}$$

★

3. STRONG MARKOV AND REFLECTION

Strong Markov

The first main result in this section, the strong Markov property (21), is due to Hunt (1956) and Blumenthal (1957). Remember that B is normalized Brownian motion on $(\Omega, \mathscr{F}, \mathscr{P})$, and $\mathscr{F}(t)$ is the σ-field spanned by $B(s)$ for $0 \leqq s \leqq t$. For each $t \geqq 0$, let $\mathscr{A}(t)$ be a sub-σ-field of $\mathscr{F}$, such that

$\mathscr{A}(t) \supset \mathscr{F}(t)$, and

$\mathscr{A}(u) \supset \mathscr{A}(v)$ when $u > v$, and

$\mathscr{A}(t)$ is independent of $\{B(t + s) - B(t): 0 \leqq s < \infty\}$.

Then $\{\mathscr{A}(t)\}$ is *admissible*. As (11c) shows, $\{\mathscr{F}(t)\}$ is admissible. If the σ-field $\mathscr{A}$ is independent of B, and $\mathscr{A}(t)$ is the σ-field generated by $\mathscr{A}$ and $\mathscr{F}(t)$, then $\{\mathscr{A}(t)\}$ is admissible. More complicated examples can be constructed. A nonnegative, extended real-valued random variable τ on $(\Omega, \mathscr{F})$ is *Markov* relative to $\mathscr{A}(t)$ iff $\{\tau < t\} \in \mathscr{A}(t)$ for all $t \geqq 0$. *Markov* without further qualification means relative to $\{\mathscr{F}(t)\}$. The *pre-τ sigma-field* $\mathscr{A}(\tau+)$ is the σ-field of all sets $A \in \mathscr{F}$ such that $A \cap \{\tau < t\} \in \mathscr{A}(t)$ for all $t \geqq 0$. For a discussion, see Section 6.6 of *MC*. The process $B(\tau + \cdot) - B(\tau)$ is the map

$$(t, \omega) \to B[\tau(\omega) + t, \omega] - B[\tau(\omega), \omega].$$

It is defined for $0 \leqq t < \infty$ and $\omega \in \{\tau < \infty\}$, and is jointly measurable: that is, with respect to the product of the Borel σ-field on $[0, \infty)$ and the relativization of $\mathscr{F}$ to $\{\tau < \infty\}$. Indeed, $(t, \omega) \to B(t, \omega)$ is continuous in t for each ω, and is measurable in ω for each t, which makes it jointly measurable. Next, $(t, \omega) \to \tau(\omega) + t$ is jointly measurable, being the sum of two jointly measurable functions. So $(t, \omega) \to (\tau(\omega) + t, \omega)$ is jointly measurable. Now $(t, \omega) \to B(\tau(\omega) + t, \omega)$ is the composition of B with the latter function.

EXAMPLE. $B(\tau)$ is measurable relative to $\mathscr{A}(\tau +)$.

PROOF. Let τ_n be the least $j/2^n > \tau$. Let y be a real number. Then

$$\{B(\tau) \geqq y \text{ and } \tau < t\} = \textstyle\bigcap_m \limsup_n A_{m,n}, \quad \text{where}$$

$$A_{m,n} = \{B(\tau_n) \geqq y - m^{-1} \text{ and } \tau_n < t\}.$$

Next,

$$\{B(\tau_n) \geqq z \text{ and } \tau_n < t\} = \textstyle\bigcup_j A_j, \quad \text{where}$$

$$A_j = \{B(j/2^n) \geqq z \text{ and } (j - 1)/2^n \leqq \tau < j/2^n\}, \quad \text{and}$$

$$j = 1, \ldots, \langle 2^n t\rangle.$$

★

(21) Theorem. *Suppose $\{\mathscr{A}(t)\}$ is admissible, and τ is Markov relative to $\{\mathscr{A}(t)\}$. Given $\{\tau < \infty\}$, the pre-τ sigma-field $\mathscr{A}(\tau +)$ is conditionally $\mathscr{P}$-independent of the process $\mathscr{B}(\tau + \cdot) - B(\tau)$, which is conditionally normalized Brownian motion.*

PROOF. Fix $0 < t_1 < \cdots < t_m < \infty$ and bounded, continuous functions $g_1, \ldots, g_m$ on the real line. Define a bounded, measurable function γ on $C[0, \infty)$ as follows:

$$\gamma(f) = g_1[f(t_1)] \cdot g_2[f(t_2)] \cdots g_m[f(t_m)] \quad \text{for } f \in C[0, \infty).$$

As m, the t's, and the g's vary, the σ-field spanned by the γ's coincides with the one spanned by the coordinates. My problem is to show

$$\int_A \gamma[B(\tau + \cdot) - B(\tau)]\, d\mathscr{P} = \mathscr{P}\{A\} \cdot \int \gamma(B)\, d\mathscr{P} \tag{22}$$

for all $A \in \mathscr{A}(\tau +)$ with $A \subset \{\tau < \infty\}$.

Let τ_n be the least $j/2^n > \tau$. I claim:

$$\int_A \gamma[B(\tau_n + \cdot) - B(\tau_n)]\, d\mathscr{P} = \mathscr{P}\{A\} \cdot \int \gamma(B)\, d\mathscr{P}. \tag{23}$$

Indeed, let $A_j = A \cap \{\tau_n = j/2^n\}$. Then $A_j \in \mathscr{A}(j/2^n)$. By admissibility and (11c),

$$\int_{A_j} \gamma[B(\tau_n + \cdot) - B(\tau_n)]\, d\mathscr{P} = \int_{A_j} \gamma[B(j2^{-n} + \cdot) - B(j2^{-n})]\, d\mathscr{P}$$

$$= \mathscr{P}\{A_j\} \cdot \int \gamma(B)\, d\mathscr{P}.$$

Sum out j to get (23). As $n \to \infty$, the left side of (23) approaches the left side of (22), by dominated convergence. ★

Theorem (21) can be restated as follows. Remember that P_x is the $\mathscr{P}$-distribution of $x + B$, and is a probability on $C[0, \infty)$. Abbreviate $P_{B(\tau)}$ for the function

$$(\omega, A) \to P_{B[\tau(\omega), \omega]}(A)$$

of pairs (ω, A) with $\omega \in \{\tau < \infty\}$ and A a measurable subset of $C[0, \infty)$.

(24) Corollary. *Suppose $\{\mathscr{A}(t)\}$ is admissible and τ is Markov relative to $\{\mathscr{A}(t)\}$. A regular conditional $\mathscr{P}$-distribution for $B(\tau + \cdot)$ given $\mathscr{A}(\tau +)$ is $P_{B(\tau)}$, on $\{\tau < \infty\}$.*

PROOF. As in (12), using (21). ★

The reflection principle

The material in this subsection can be found in (Lévy, 1954, 1965). It will be used only in Section 4. Let τ be Markov. Define a new process Y as B *reflected* at τ:

if $\tau(\omega) = \infty$, then $Y(t, \omega) = B(t, \omega)$ for all $t \geqq 0$;

if $\tau(\omega) < \infty$, then $Y(t, \omega) = B(t, \omega)$ for all t with $0 \leqq t \leqq \tau(\omega)$;

if $\tau(\omega) < \infty$, then $Y(t, \omega) = 2B[\tau(\omega), \omega] - B(t, \omega)$ for $t \geqq \tau(\omega)$.

Geometrically, Y agrees with B up to time τ, and is then B reflected about the horizontal line through $(\tau, B(\tau))$. Namely, the distances of $Y(t)$ and $B(t)$ from $B(\tau)$ are equal and opposite:

$$Y(t) - B(\tau) = -[B(t) - B(\tau)].$$

This was illustrated in Figure 1. The reflection principle is

(25) Theorem. *Y is normalized Brownian motion.*

PROOF. Clearly, Y has continuous sample functions and starts from 0. Let A be a measurable subset of $C[0, \infty)$. Of course,

$$\mathscr{P}\{Y \in A \text{ and } \tau = \infty\} = \mathscr{P}\{B \in A \text{ and } \tau = \infty\}.$$

The challenge is

$$\mathscr{P}\{Y \in A | \tau < \infty\} = \mathscr{P}\{B \in A | \tau < \infty\}.$$

For $f \in C[0, \infty)$, and $g \in C[0, \infty)$ with $g(0) = 0$, and real $T \geqq 0$, define $\langle f, T, g\rangle \in C[0, \infty)$ by

$$\begin{aligned} \langle f, T, g\rangle(t) &= f(t) && \text{for } 0 \leqq t \leqq T \\ &= f(t) + g(t - T) && \text{for } T \leqq t < \infty. \end{aligned}$$

You should check that $(f, T, g) \to \langle f, T, g\rangle$ is product measurable.

Suppose $\tau(\omega) < \infty$. Define $f(\cdot, \omega) \in C[0, \infty)$ by:

$$\begin{aligned} f(t, \omega) &= B(t, \omega) && \text{for } 0 \leqq t \leqq \tau(\omega) \\ &= B[\tau(\omega), \omega] && \text{for } \tau(\omega) \leqq t < \infty. \end{aligned}$$

Define $g(\cdot, \omega) \in C[0, \infty)$ by:

$$g(t, \omega) = B[\tau(\omega) + t, \omega] - B[\tau(\omega), \omega].$$

You should check that τ and f are measurable with respect to the pre-τ sigma field.

With respect to $\mathscr{P}$, given $\tau < \infty$: by strong Markov (21),

g is independent of (f, τ), and has distribution P_0;

let $h = -g$; now symmetry (11a) implies

h is independent of (f, τ), and has distribution P_0;

consequently,

the distribution of (f, τ, g) coincides with the distribution of (f, τ, h);

and

the distribution of $\langle f, \tau, g\rangle$ coincides with the distribution of $\langle f, \tau, h\rangle$.

But $\langle f, \tau, g\rangle = B$ and $\langle f, \tau, h\rangle = Y$ on $\{\tau < \infty\}$. ★

The next result improves (16a). To state it, let

$$M(t) = \max\{B(s): 0 \leqq s \leqq t\}.$$

(26) Theorem. *Let $y \geqq 0$ and $b \geqq 0$. Then*

$$\mathscr{P}\{B(t) < b - y \text{ and } M(t) \geqq b\} = \mathscr{P}\{B(t) > b + y\}.$$

PROOF. Let τ be the least s with $B(s) = b$: rely on (18). Confirm that $\{M(t) \geqq b\} = \{\tau \leqq t\}$. Now

$$M(t) = \sup\{B(r): 0 \leqq r \leqq t \text{ and } r \text{ is rational}\}$$

is $\mathscr{F}(t)$-measurable: where $\mathscr{F}(t)$ is the σ-field spanned by $B(s)$ for $s \leqq t$. Thus

$$\{\tau < t\} = \bigcup_n \left\{\tau \leqq t - \frac{1}{n}\right\} \in \mathscr{F}(t),$$

and τ is Markov. I claim:

(27) $$\mathscr{P}\{\tau \leqq t \text{ and } B(t) > b + y\} = \mathscr{P}\{\tau \leqq t \text{ and } B(t) < b - y\}.$$

Indeed, let Y be B reflected at τ, as for (25). Let σ be the least t with $Y(t) = b$. Since Y is distributed like B, and σ is the same measurable function of Y that τ is of B,

$$\mathscr{P}\{\sigma \leqq t \text{ and } Y(t) < b - y\} = \mathscr{P}\{\tau \leqq t \text{ and } B(t) < b - y\}.$$

But $\sigma = \tau$ and

$$\{\sigma \leqq t \text{ and } Y(t) < b - y\} = \{\tau \leqq t \text{ and } B(t) > b + y\},$$

so

(28) $\mathscr{P}\{\sigma \leqq t \text{ and } Y(t) < b - y\} = \mathscr{P}\{\tau \leqq t \text{ and } B(t) > b + y\}.$

This proves (27).

Finally,

$$\{B(t) \geqq b\} \subset \{\tau \leqq t\}. \qquad \bigstar$$

(29) Corollary. *For* $b \geqq 0$ *and* $t > 0$,

$$\mathscr{P}\{M(t) \geqq b\} = 2\mathscr{P}\{B(t) \geqq b\}.$$

PROOF. Use (26), and remember $\mathscr{P}\{B(t) = b\} = 0$:

$$\begin{aligned}\mathscr{P}\{M(t) \geqq b\} &= \mathscr{P}\{B(t) < b \leqq M(t)\} + \mathscr{P}\{B(t) \geqq b\}\\ &= \mathscr{P}\{B(t) > b\} + \mathscr{P}\{B(t) \geqq b\}\\ &= 2\mathscr{P}\{B(t) \geqq b\}. \qquad \bigstar\end{aligned}$$

(30) Corollary. *The joint distribution of* $B(t)$ *and* $M(t)$ *is absolutely continuous with respect to planar Lebesgue measure, for* $t > 0$. *The joint density* $f(x, b)$ *is given by the formula*

$$\begin{aligned}f(x, b) &= 0 \quad \textit{when } x > b \textit{ or } b < 0\\ &= \sqrt{\frac{2}{\pi}}\,\frac{2b - x}{t^{3/2}}\, e^{-\frac{(2b - x)^2}{2t}} \quad \textit{when } b \geqq 0 \textit{ and } b \geqq x.\end{aligned}$$

PROOF. Put $x = b - y$ in (26), and then differentiate. ★

The next main result† is (33), to which (31) is preliminary. For these results, let τ_x be the least t with $B(t) = x$, relying on (18). Let $a < 0 < b$. Let A be $\{\tau_a < \tau_b\}$ and $C = \Omega \setminus A = \{\tau_b < \tau_a\}$. Fix $t > 0$. Let $A^* = \{\tau_a \leqq t\}$ and $C^* = \{\tau_b \leqq t\}$. Let $r_y(x)$ be the reflection of x around y, namely $2y - x$. If H is a Borel subset of $(-\infty, \infty)$, let

$$r_y(H) = \{r_y(h) : h \in H\} \quad \text{and} \quad \mathbf{H} = \{\omega : \omega \in \Omega \text{ and } B(t, \omega) \in H\}.$$

Consult Figure 4 before tackling (31) and (33).

(31) Lemma. **(a)** *If* $H \subset (-\infty, a]$, *then*

$$\mathscr{P}\{C \cap \mathbf{H}\} = \mathscr{P}\{\mathbf{r_b H}\} - \mathscr{P}\{A \cap \mathbf{r_b H}\}.$$

(b) *If* $H \subset [b, \infty)$, *then*

$$\mathscr{P}\{A \cap \mathbf{H}\} = \mathscr{P}\{\mathbf{r_a H}\} - \mathscr{P}\{C \cap \mathbf{r_a H}\}.$$

† On a first reading of the book, skip to Section 4.

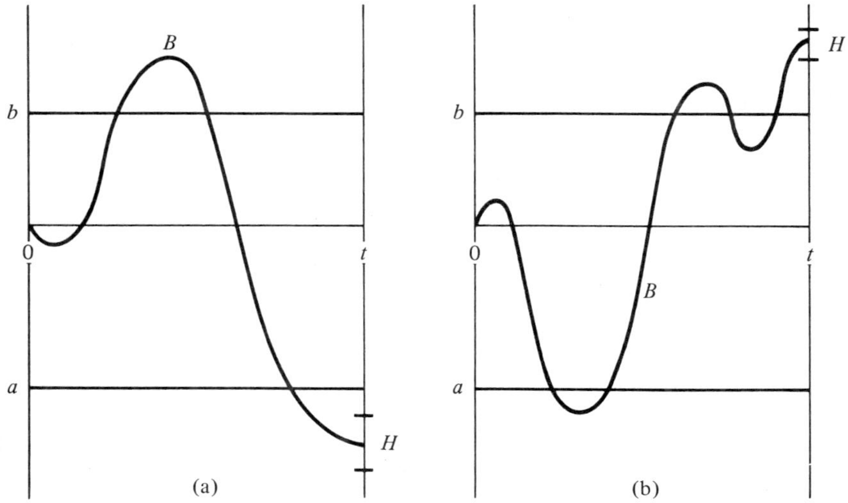

Figure 4.

Proof. I will only argue (a). Clearly, $\mathbf{H} \subset A^*$, so $C \cap \mathbf{H} \subset B^*$. Moreover, $r_b H \subset [b, \infty)$, so $\mathbf{r_b H} \subset B^*$. Use the reflection principle (25) with τ_b for τ, to check

$$\mathscr{P}\{C \cap \mathbf{H}\} = \mathscr{P}\{C \cap \mathbf{r_b H}\}.$$

Put $\Omega \setminus A$ for C on the right. ★

Let $m(t) = \min\{B(s): 0 \leqq s \leqq t\}$. Recall $M(t) = \max\{B(s): 0 \leqq s \leqq t\}$. Let

$$(32) \qquad k(y) = k(a, b, t : y) = \frac{1}{\sqrt{2\pi t}} \Sigma_{n=-\infty}^{\infty} \left[e^{-\frac{(y-2nc)^2}{2t}} - e^{-\frac{(y-2a+2nc)^2}{2t}} \right]$$

for $t > 0$ and $a < 0 < b$ and $c = b - a$.

(33) Theorem. *Let J be a Borel subset of $[a, b]$, with $a < 0 < b$, and let $c = b - a$. Define k by (32). Then*

$$\mathscr{P}\{a < m(t) \leqq M(t) < b \text{ and } B(t) \in J\} = \int_J k(y)\, dy.$$

Proof. Continue with the notation for (31). First, the required probability is

$$\mathscr{P}\{\mathbf{J}\} - \mathscr{P}\{A \cap A^* \cap \mathbf{J}\} - \mathscr{P}\{C \cap C^* \cap \mathbf{J}\}:$$

because $m(t) \leqq a$ or $M(t) \geqq b$ means that the path hits a or b on or before time t, and one or the other has to be hit first; $A \cap A^*$ is the event that the path hits a on or before time t, and before hitting b; while $C \cap C^*$ is the event that the path hits b on or before time t, and before hitting a. Now the second and third terms are evaluated in a similar way; only the second, T_2, will be done. By the reflection principle (25),

$$T_2 = \mathscr{P}\{A \cap A^* \cap \mathbf{J}\} = \mathscr{P}\{A \cap A^* \cap \mathbf{r_a J}\}.$$

Now $r_a J \subset (-\infty, a]$, so $\mathbf{r_a J} \subset A^*$. Thus

$$T_2 = \mathscr{P}\{A \cap \mathbf{r_a J}\} = \mathscr{P}\{\mathbf{r_a J}\} - \mathscr{P}\{C \cap \mathbf{r_a J}\}.$$

Using (31a) and (31b) alternately,

$$T_2 = \mathscr{P}\{\mathbf{r_a J}\} - \mathscr{P}\{\mathbf{r_b r_a J}\} + \mathscr{P}\{\mathbf{r_a r_b r_a J}\} - \cdots,$$

where the series converges absolutely by an easy estimate: with $c = b - a$,

$$(r_a r_b)^n(x) = x - 2nc \quad \text{and} \quad (r_b r_a)^n(x) = x + 2nc,$$

so the intervals $r_a J$, $r_b r_a J$, $r_a r_b r_a J, \ldots$ recede from the origin quite rapidly. The nth negative term in T_2—which is positive overall because T_2 is taken with negative sign—is

$$\begin{aligned}\mathscr{P}\{B(t) \in (r_b r_a)^n J\} &= \mathscr{P}\{B(t) - 2nc \in J\} \\ &= \int_J \frac{1}{\sqrt{2\pi t}} e^{-\frac{(y+2nc)^2}{2t}}\, dy.\end{aligned}$$

The nth positive term in T_2 can be computed in the same way. The rest is housekeeping. ★

An argument which is similar but easier proves

(34) Proposition. *Let b and t be positive.*

$$\begin{aligned}&\mathscr{P}\{\max_{0 \leqq s \leqq t} |B(s)| \leqq b\} \\ &\quad = \mathscr{P}\{0 \leqq |B(t)| \leqq b\} - \mathscr{P}\{b \leqq |B(t)| \leqq 3b\} \\ &\qquad + \mathscr{P}\{3b \leqq |B(t)| \leqq 5b\} - \mathscr{P}\{5b \leqq |B(t)| \leqq 7b\} \\ &\qquad + \cdots.\end{aligned}$$

This sharpens (16c), in a disguised way: subtract both sides from 1, and confirm that

$$\begin{aligned}\mathscr{P}\{|B(t)| \geqq b\} \geqq \mathscr{P}\{b \leqq |B(t)| \leqq 3b\} &- \mathscr{P}\{3b \leqq |B(t)| \leqq 5b\} \\ &+ \mathscr{P}\{5b \leqq |B(t)| \leqq 7b\} - \cdots.\end{aligned}$$

The next result will not be used later; the proof is only sketched.

(35) Proposition. *Define k by (32). If $t > 0$ and $a < 0 < b$, then k is strictly positive on (a, b).*

PROOF. Dominated convergence implies that k is a continuous function of y. So (33) shows

(36) $k(a, b, t : y)$ is the limit as $\varepsilon \downarrow 0$ of

$$\frac{1}{2\varepsilon}\mathscr{P}\{a < m(t) \leqq M(t) < b \text{ and } y - \varepsilon < B(t) < y + \varepsilon\}$$

for $a < 0 < b$ and $t > 0$. Fix $t > 0$. By symmetry (11a), I only have to get k strictly positive on $[0, b)$. Fix y and b with $b > y \geqq 0$. Let

$$f(a) = k(a, b, t : y)$$

for $a < 0$. By (32) and analysis, $f(0-) = 0$. And $f(a)$ is nondecreasing as a decreases, by (36). By dominated convergence,

$$\lim_{a\downarrow -\infty} f(a) = \frac{1}{\sqrt{2\pi t}}\left[e^{-\frac{y^2}{2t}} - e^{-\frac{(y-2b)^2}{2t}}\right];$$

which is positive by algebra. Moreover, $f(a)$ is an analytic function of $a \in (-\infty, 0)$, even of complex a with negative real part: use (3.77 and 3.79). Since f is not constant on $(-\infty, 0)$, it is constant in no subinterval (3.80). That is, $f(a)$ increases strictly as a decreases, and is therefore strictly positive on $(-\infty, 0)$. ★

The paths are dense

The next result is weaker than (35), but it can be proved without recourse to (25) or complex variables, and it is enough for (38).

(37) Lemma. *Let $a < 0 < b$ and $t > 0$. Let J be a subinterval of $[a, b]$. Suppose J has a nonempty interior. Then*

$$\mathscr{P}\{a < m(t) \leqq M(t) < b \text{ and } B(t) \in J\} > 0.$$

PROOF. *Case* 1: $b = -a$ *and* $J = [a, b]$. Let τ be the least t such that $B(t) = \pm b/2$. Find $\delta > 0$ with $\mathscr{P}\{\tau \geqq \delta\} > 0$. As symmetry (11a) shows,

$$\mathscr{P}\{\tau \geqq \delta \text{ and } B(\tau) = b/2\} = \tfrac{1}{2}\mathscr{P}\{\tau \geqq \delta\}.$$

Choose even n so large that $n\delta > t$. Let $\tau_0 = 0$ and let τ_{j+1} be the least $t > \tau_j$ with $|B(t) - B(\tau_j)| = b/2$, so $\tau_1 = \tau$. The event

$$\tau_1 \geqq \delta.\ \tau_2 - \tau_1 \geqq \delta, \ldots, \tau_n - \tau_{n-1} \geqq \delta$$

$$B(\tau_1) = \frac{b}{2}, B(\tau_2) = 0, B(\tau_3) = \frac{b}{2}, \ldots, B(\tau_n) = 0$$

is a subevent of $\{\max_{0 \leqq s \leqq t} |B(s)| < b\}$, and has probability $[\frac{1}{2}\mathscr{P}\{\tau \geqq \delta\}]^n > 0$, by the strong Markov property (21).

The general case. Let μ be the midpoint of J. Choose $\varepsilon > 0$ with $(\mu - \varepsilon, \mu + \varepsilon) \subset J$. Without real loss (11a), suppose $\mu < 0$. Let τ be the least s with $B(s) = \mu$ or b. Let

$$D = \{\tau < t \text{ and } B(\tau) = \mu\}$$

$$G = \{D \text{ and } |B(s) - \mu| < \varepsilon \text{ for } \tau \leqq s \leqq t\}.$$

I will argue that $\mathscr{P}\{G\} > 0$, which does (37).

The first step is to show that $\mathscr{P}\{D\} > 0$. Use (26):

$$\begin{aligned}
\mathscr{P}\{B(t) < \mu \text{ and } M(t) < b\} &= \mathscr{P}\{B(t) < \mu\} - \mathscr{P}\{B(t) < \mu \text{ and } M(t) \geqq b\} \\
&= \mathscr{P}\{B(t) < \mu\} - \mathscr{P}\{B(t) > 2b - \mu\} \\
&= \mathscr{P}\{\mu - 2b \leqq B(t) < \mu\} \\
&> 0.
\end{aligned}$$

Clearly, $D \supset \{B(t) < \mu \text{ and } M(t) < b\}$, forcing $\mathscr{P}\{D\} > 0$.

To continue on $\mathscr{P}\{G\}$, consider the cartesian product

$$\Omega \times C[0, \infty),$$

endowed with the product of $\mathscr{F}(\tau+)$ and the usual σ-field on $C[0, \infty)$. Let G^* be the set of pairs (ω, f) with $\omega \in D$ and

$$|f(s)| < \varepsilon \quad \text{for } 0 \leqq s \leqq t - \tau(\omega).$$

Of course, G^* is product measurable, because $D \in \mathscr{F}(\tau+)$ and τ is $\mathscr{F}(\tau+)$-measurable. The ω-section $G^*(\omega)$ of G^* is empty for $\omega \notin D$; and for $\omega \in D$,

$$G^*(\omega) = \{f : f \in C[0, \infty) \text{ and } |f(s)| < \varepsilon \text{ for } 0 \leqq s \leqq t - \tau(\omega)\}.$$

Remember that P_0 is the distribution of B. Let

$$\theta(u) = P_0\{f : f \in C[0, \infty) \text{ and } |f(s)| < \varepsilon \text{ for } 0 \leqq s \leqq u\}.$$

By case 1, for $\omega \in D$,

$$P_0\{G^*(\omega)\} = \theta(t - \tau(\omega)) > 0.$$

Let

$$Y(t, \omega) = B[\tau(\omega) + t, \omega] - B[\tau(\omega), \omega].$$

Then

$$G = \{\omega : \omega \in \Omega \text{ and } (\omega, Y(\cdot, \omega)) \in G^*\}.$$

As strong Markov (21) implies, Y is independent of $\mathscr{F}(\tau+)$ and has distribution P_0. By Fubini,

$$\begin{aligned}\mathscr{P}\{G\} &= \int P_0\{G^*(\omega)\}\,\mathscr{P}(d\omega)\\ &= \int_D \theta(t - \tau(\omega))\,\mathscr{P}(d\omega)\\ &> 0.\end{aligned}$$

★

The next result states that Brownian motion paths are dense. More precisely, let f be a continuous function on $[0, 1]$, with $f(0) = 0$. Let $\varepsilon > 0$.

(38) Theorem. $\mathscr{P}[|B(t) - f(t)| \leqq \varepsilon \textit{ for } 0 \leqq t \leqq 1] > 0$.

PROOF. Choose n so large that $|x - y| \leqq 1/n$ implies $|f(x) - f(y)| \leqq \varepsilon/3$. Choose positive $\varepsilon_1, \ldots, \varepsilon_n$ with $\varepsilon_1 + \cdots + \varepsilon_n \leqq \varepsilon/3$. For $i = 1, \ldots, n$, let A_i be the event that

$$\varepsilon_i \geqq \left|\left[B\left(\frac{i}{n}\right) - B\left(\frac{i-1}{n}\right)\right] - \left[f\left(\frac{i}{n}\right) - f\left(\frac{i-1}{n}\right)\right]\right|$$

and

$$\frac{\varepsilon}{3} \geqq \max\left\{\left|B(t) - B\left(\frac{i-1}{n}\right)\right| : \frac{i-1}{n} \leqq t \leqq \frac{i}{n}\right\}.$$

Now $|f(i/n) - f((i-1)/n)| \leqq \varepsilon/3$ and $\varepsilon_i > 0$, so (37) and (11c) imply $\mathscr{P}\{A_i\} > 0$. Using (11c) again, the A_i are independent; so $\mathscr{P}\{\bigcap_{i=1}^n A_i\} > 0$. But $\omega \in \bigcap_{i=1}^n A_i$ and $t \in [0, 1]$ imply $|B(t, \omega) - f(t)| \leqq \varepsilon$; indeed, find $i = 1, \ldots, n$ with $(i-1)/n \leqq t \leqq i/n$. Then

$$\left|B(t, \omega) - B\left(\frac{i-1}{n}, \omega\right)\right| \leqq \frac{\varepsilon}{3}$$

and

$$\left|B\left(\frac{i-1}{n}, \omega\right) - f\left(\frac{i-1}{n}\right)\right| \leqq \varepsilon_1 + \cdots + \varepsilon_{i-1} \leqq \frac{\varepsilon}{3}$$

and

$$\left|f(t) - f\left(\frac{i-1}{n}\right)\right| \leqq \frac{\varepsilon}{3}.$$

★

Markov processes associated with Brownian motion

The next main result is (42); here is a preliminary discussion. A *Markov semigroup* K is a function $K(t, x, A)$ of three variables: t is a nonnegative real number; x is a real number; and A is a Borel subset of $(-\infty, \infty)$. Moreover, for each t and A, the function $K(t, \cdot, A)$ is measurable; and for each t and x, the function $K(t, x, \cdot)$ is a probability on the Borel subsets of $(-\infty, \infty)$. Finally, for all t, s, x and A:

$$K(t+s, x, A) = \int_{-\infty}^{\infty} K(t, y, A)\, K(s, x, dy).$$

Remember that ξ is the coordinate process on $C[0, \infty)$:

$$\xi(t, f) = f(t)$$

for $0 \leqq t < \infty$ and $f \in C[0, \infty)$.

A probability P on $C[0, \infty)$ is *Markov starting from* x_0 *with stationary transitions* K iff $0 < t_1 < t_2 < \cdots$ and $A_1, A_2, \ldots$ Borel imply

$$P\{\xi(0) = x_0\} = 1$$

$$P\{\xi(0) = x_0 \text{ and } \xi(t_1) \in A_1\} = K(t_1, x_0, A_1)$$

$$P\{\xi(0) = x_0 \text{ and } \xi(t_1) \in A_1 \text{ and } \xi(t_2) \in A_2\}$$

$$= \int_{x_1 \in A_1} K(t_2 - t_1, x_1, A_2)\, K(t_1, x_0, dx_1),$$

and so on. Formally,

$$P\{\xi(0) = x_0\} = 1;$$

and a regular conditional P-distribution for $\xi(t_{n+1})$ given $\xi(t_1), \ldots, \xi(t_n)$ is

$$K(t_{n+1} - t_n, \xi(t_n), \cdot):$$

in other terms,

$$P\{F \text{ and } \xi(t+s) \in A\} = \int_F K(t, \xi(s), A)\, dP$$

for all F in the σ-field generated by $\xi(u)$ with $0 \leqq u \leqq s$.

A process with continuous sample functions is Markov starting from x_0 with stationary transitions K iff its distribution is.

(39) Example. Normalized Brownian motion is Markov starting from 0 with stationary transitions K, where $K(t, x, \cdot)$ is the normal distribution with mean x and variance t.

PROOF. Use definition (1). ★

To state (40), and for future use, let $|N(\mu, \sigma^2)|$ be the distribution of $|Z|$, where Z is $N(\mu, \sigma^2)$. Of course,

$$|N(\mu, \sigma^2)| = |N(-\mu, \sigma^2)| = |N(|\mu|, \sigma^2)|.$$

Let $K_1(t, x, \cdot)$ be the $\mathscr{P}$-distribution of $|x + B(t)|$: namely, $N|(x, t)|$.

(40) Example. K_1 is a Markov semigroup. The process $|B|$ is Markov starting from 0, with stationary transitions K_1.

PROOF. Fix a Borel subset A of the line. Write $|N(\mu, \sigma^2)|(A)$ for the $|N(\mu, \sigma^2)|$-measure of A. Fix nonnegative s and t. Remember that $\mathscr{F}(s)$ is the σ-field generated by $B(u)$ for $0 \leqq u \leqq s$. Fix $F \in \mathscr{F}(s)$. Use Fubini and (11c):

$$\begin{aligned}
&\mathscr{P}\{F \text{ and } |x + B(t + s)| \in A\} \\
&\qquad = \mathscr{P}\{F \text{ and } |x + B(s) + B(t + s) - B(s)| \in A\} \\
&\qquad = \int_F |N(x + B(s), t)|(A)\, d\mathscr{P} \\
&\qquad = \int_F |N(|x + B(s)|, t)|(A)\, d\mathscr{P} \\
&\qquad = \int_F K_1(t, |x + B(s)|, A)\, d\mathscr{P}.
\end{aligned}$$

This is even truer if you confine F to the smaller σ-field generated by $|x + B(u)|$ for $0 \leqq u \leqq s$; then it's the Markov property. If you put $F = \Omega$, you get the semigroup property. ★

NOTE. $|B|$ is called *normalized Brownian motion with reflecting barrier at* 0.

(41) Lemma. *Let P and Q be two probabilities on $C[0, \infty)$, both Markov with same stationary transitions and starting value. Then $P = Q$.*

PROOF. Use (9). ★

Abbreviate $a \vee b = \max\{a, b\}$. Recall $M(t) = \max_{0 \leqq s \leqq t} B(s)$. Define $Y(t) = M(t) - B(t)$.

Side comment. M is not Markov.

(42) Theorem. *The $\mathscr{P}$-distribution of Y coincides with the $\mathscr{P}$-distribution of $|B|$.*

PROOF. The program is to use (41). Both $|B|$ and Y have continuous sample functions starting from 0. And $|B|$ is Markov with stationary transitions K_1. What I have to argue is that Y is Markov with the same transitions.

Fix a nonnegative real number s. Let

$$\hat{B}(u) = B(s + u) - B(s) \quad \text{for } 0 \leqq u < \infty$$

$$\hat{M}(t) = \max\{\hat{B}(u) : 0 \leqq u \leqq t\}.$$

I say

$$Y(s+t) = [Y(s) \vee \hat{M}(t)] - \hat{B}(t). \tag{43}$$

Indeed,

$$B(s+t) = B(s) + \hat{B}(t)$$

$$B(s) + \hat{M}(t) = \max\{B(s+u): 0 \leqq u \leqq t\}$$

$$M(s+t) = M(s) \vee [B(s) + \hat{M}(t)].$$

So

$$\begin{aligned} Y(s+t) &= M(s+t) - B(s+t) \\ &= \{M(s) \vee [B(s) + \hat{M}(t)]\} - B(s) - \hat{B}(t) \\ &= \{[M(s) - B(s)] \vee \hat{M}(t)\} - \hat{B}(t) \\ &= [Y(s) \vee \hat{M}(t)] - \hat{B}(t). \end{aligned}$$

This proves (43).

Let $y \geqq 0$. Let $Q(t, y, \cdot)$ be the $\mathscr{P}$-distribution of

$$[y \vee M(t)] - B(t).$$

Remember that $\mathscr{F}(s)$ is the σ-field spanned by $B(u)$ for $0 \leqq u \leqq s$. As (11c) states, $\hat{B}$ is independent of $\mathscr{F}(s)$, and its distribution is P_0. So, the pair $(\hat{M}(t), \hat{B}(t))$ is independent of $\mathscr{F}(s)$, and is distributed like $(M(t), B(t))$. Of course, $Y(s)$ is $\mathscr{F}(s)$-measurable. So $\mathscr{F}(s)$ includes the σ-field $\mathscr{F}_y(s)$ generated by $Y(u)$ with $0 \leqq u \leqq s$. Now use Fubini and (43):

$$\mathscr{P}\{F \text{ and } Y(s+t) \in A\} = \int_F Q(t, Y(s), A)\, d\mathscr{P} \quad \text{for } F \in \mathscr{F}_y(s). \tag{44}$$

The last problem is showing that

$$[y \vee M(t)] - B(t) \quad \text{is distributed like} \quad |y + B(t)|. \tag{45}$$

This would clinch the proof: (44) and (45) together make Y Markov with stationary transitions K_1, which is all I needed. I will now argue (45). Let

$$Z(u) = B(t-u) - B(t) \quad \text{for } 0 \leqq u \leqq t.$$

The distribution of $\{Z(u): 0 \leqq u \leqq t\}$ coincides with that of $\{B(u): 0 \leqq u \leqq t\}$ by (11d). Now

$$\begin{aligned} [y \vee M(t)] - B(t) &= [y - B(t)] \vee [M(t) - B(t)] \\ &= [y + Z(t)] \vee \max\{Z(u): 0 \leqq u \leqq t\}. \end{aligned}$$

Thus, (45) is equivalent to

$$[y + B(t)] \vee M(t) \quad \text{is distributed like} \quad |y + B(t)|. \tag{46}$$

Fix $a > 0$. Then (26) and symmetry show

$$\begin{aligned}\mathscr{P}\{[y + B(t)] \vee M(t) \geqq a\} &= \mathscr{P}\{y + B(t) \geqq a\} + \mathscr{P}\{y + B(t) < a \leqq M(t)\}\\ &= \mathscr{P}\{y + B(t) \geqq a\} + \mathscr{P}\{B(t) > a + y\}\\ &= \mathscr{P}\{y + B(t) \geqq a\} + \mathscr{P}\{B(t) \leqq -a - y\}\\ &= \mathscr{P}\{y + B(t) \geqq a\} + \mathscr{P}\{y + B(t) \leqq -a\}\\ &= \mathscr{P}\{|y + B(t)| \geqq a\}.\end{aligned}$$ ★

To end the section, here are two more Markov processes associated with Brownian motion. To state the first example (47a), let $b > 0$. Let τ_b be the least t with $B(t) = b$. Let

$$\begin{aligned}Y(t) &= B(t) \quad \text{for } t \leqq \tau_b\\ &= b \qquad \text{for } t > \tau_b.\end{aligned}$$

Then Y is *normalized Brownian motion with absorbing barrier at b*. Define $K_2(t, x, A)$ as follows. If $t = 0$, or $t > 0$ but $x \geqq b$, then $K_2(t, x, \cdot)$ is point mass at x. Suppose $t > 0$ and $x < b$. Then $K_2(t, x, \cdot)$ is a measure on $(-\infty, b]$, whose retraction to $(-\infty, b)$ is absolutely continuous with respect to Lebesgue measure, having density

$$y \to k_2(t, x, y) = \frac{1}{\sqrt{2\pi t}}\left[e^{-\frac{(y-x)^2}{2t}} - e^{-\frac{(2b-y-x)^2}{2t}}\right].$$

Moreover,

$$K_2(t, x, \{b\}) = \mathscr{P}\{M(t) \geqq b - x\} = 2\mathscr{P}\{B(t) \geqq b - x\},$$

the last equality coming from (29). Finally, $K_3(t, x, \cdot)$ vanishes on (b, ∞).

(47a) Example. K_2 is a Markov semigroup. Normalized Brownian motion with absorbing barrier at b is Markov starting from 0, with stationary transitions K_2.

PROOF. As in (40), using (26). ★

To state the second example (47b), remember that τ_x is the least t with $B(t) = x$. Let $a < 0 < b$. Let τ be the least t with $B(t) = a$ or b. Let

$$\begin{aligned}Y(t) &= B(t) \quad \text{for } t \leqq \tau\\ &= B(\tau) \quad \text{for } t \geqq \tau.\end{aligned}$$

Then Y is *normalized Brownian motion with absorbing barriers at a and b*. Define $K_3(t, x, A)$ as follows. For $t = 0$ or $t > 0$ but $x \notin (a, b)$, let $K_3(t, x, \cdot)$

be point mass 1 at x. Suppose $t > 0$ and $a < x < b$. Then $K_3(t, x, \cdot)$ is a measure on $(-\infty, \infty)$. Its retraction to (a, b) is absolutely continuous with respect to Lebesgue measure, having density

$$y \to k(a - x, b - x, t : y - x),$$

where k was defined by (32). Moreover,

$$K_3(t, x, \{a\}) = \mathscr{P}\{\tau_{a-x} \leqq t \text{ and } \tau_{a-x} < \tau_{b-x}\}$$

$$K_3(t, x, \{b\}) = \mathscr{P}\{\tau_{b-x} \leqq t \text{ and } \tau_{b-x} < \tau_{a-x}\}.$$

These two expressions were computed in the course of proving (33). Finally, $K_3(t, x, \cdot)$ vanishes outside $[a, b]$.

(47b) Example. K_3 is a Markov semigroup. Normalized Brownian motion with absorbing barriers at a and b is Markov starting from 0, with stationary transitions K_3.

PROOF. As in (40), using (33). ★

4. SAMPLE FUNCTION PROPERTIES

This section contains some results on the properties of Brownian motion sample paths. Except for (57–59), those are not used elsewhere in the book. This material, and many other striking results, can be found in (Lévy, 1954, 1965). Other references are (Doob, 1953) and (Itô and McKean, 1965).

Local maxima

(48) Theorem. *For almost all ω, the function $B(\cdot, \omega)$ is monotone in no interval.*

PROOF. Any nonempty interval includes a rational interval, that is, an interval with rational endpoints. There are only countably many such intervals. Consequently, it is enough to show that, for any given interval, $B(\cdot, \omega)$ is monotone on that interval for almost no ω. By (11), it is enough to show that $B(\cdot, \omega)$ is nondecreasing on $[0, 1]$ for almost no ω. Let

$$A_n = \left\{B\left(\frac{i+1}{n}\right) \geqq B\left(\frac{i}{n}\right) \text{ for } i = 0, \ldots, n-1\right\}$$

$$= \left\{B\left(\frac{i+1}{n}\right) - B\left(\frac{i}{n}\right) \geqq 0 \text{ for } i = 0, \ldots, n-1\right\}.$$

The differences $B\left(\frac{i+1}{n}\right) - B\left(\frac{i}{n}\right)$ are independent $N\left(0, \frac{1}{n}\right)$, so

$$\mathscr{P}(A_n) = 1/2^n \to 0.$$

But $\bigcap_n A_n$ is the set where B is nondecreasing on $[0, 1]$. ★

DEFINITION. *f has a* local maximum *at t iff there is an $\varepsilon > 0$ such that*

$$f(s) \leqq f(t) \quad \textit{for } t - \varepsilon \leqq s \leqq t + \varepsilon.$$

f has a strict local maximum *at t if there is an $\varepsilon > 0$ such that*

$$f(s) < f(t) \quad \textit{for } 0 < |t - s| \leqq \varepsilon.$$

(49) Lemma. *Let f be a continuous function on* $[0, 1]$, *monotone in no interval. Then f has a local maximum in* $[0, 1]$.

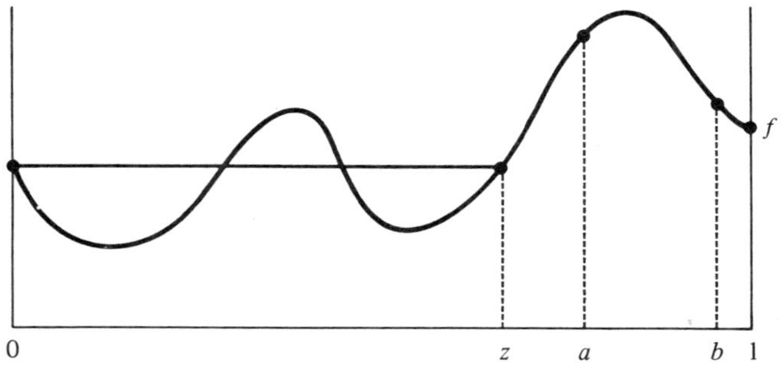

Figure 5.

PROOF. *Case 1:* $f(1) > f(0)$. See Figure 5. Let z be the largest x with $f(x) = f(0)$. Since f is not monotone in $[z, 1]$, there are a, b with $z < a < b < 1$ and $f(a) > f(b)$. Since $f(a) > f(z)$ by construction, f has a local maximum in (z, b).

Case 2: $f(1) < f(0)$. Use Case 1 on $g(x) = f(1 - x)$.

Case 3: $f(1) = f(0)$. Find $c \in [0, 1]$ with $f(c) \neq 0$. Use Case 1 or 2 on $[0, c]$. ★

(50) Corollary. *The local maxima of f are dense in* $[0, 1]$.

(51) Corollary. *For almost all ω, the set of local maxima of $B(\cdot, \omega)$ is dense.*

(52) Proposition. *For almost all ω, each local maximum of $B(\cdot, \omega)$ is strict.*

PROOF. For an interval I, let

$$M(I) = \max\{B(t) : t \in I\}.$$

Let $I = [a, b]$ and $J = [c, d]$, with

$$0 \leqq a < b < c < d.$$

Let

$$X = B(c) - B(b)$$

$$Y = \max\{B(c + t) - B(c) : 0 \leqq t \leqq d - c\}.$$

So

$$M(J) = B(b) + X + Y$$

$$\{M(J) \neq M(I)\} = \{X \neq M(I) - B(b) - Y\}.$$

I remind you that $\mathscr{F}(b)$ is the σ-field spanned by $B(t)$ for $0 \leqq t \leqq b$. Use (11c) twice: $\mathscr{F}(b)$, X, and Y are independent: X is normal with mean 0 and variance $c - b$, so $\mathscr{P}\{X \neq x\} = 1$ for any x. Clearly, $B(b)$ and $M(I)$ are $\mathscr{F}(b)$-measurable:

$$M(I) = \sup\{B(r) : r \in I \text{ and } r \text{ is rational}\}.$$

Use Fubini:

$$\mathscr{P}\{M(J) \neq M(I)\} = \int \mathscr{P}\{X \neq M(I, \omega) - B(b, \omega) - Y(\omega)\}\, \mathscr{P}(d\omega) = 1.$$

Therefore,

$$\mathscr{P}\{M(J) \neq M(I)\} = 1.$$

The intersection of $\{M(J) \neq M(I)\}$ for rational a, b, c, d is a subset of the set of ω such that each local maximum of $B(\cdot, \omega)$ is strict. ★

(53) Measurability. $B(\cdot, \omega)$ has a local maximum in $[0, 1]$ iff there are rational numbers a, b, c with $0 \leqq a < b < c \leqq 1$ and $B(b, \omega)$ no smaller than $B(a, \omega)$ or $B(c, \omega)$. The set of ω such that $B(\cdot, \omega)$ has a local maximum in $[0, 1]$ is therefore measurable.

For rational numbers $a < b$, let

$$M(a, b, \omega) = \sup\{B(r, \omega) : a \leqq r \leqq b \text{ and } r \text{ is rational}\}.$$

Then $B(\cdot, \omega)$ has a nonstrict local maximum iff there are rational a, b, c, d such that $a < b < c < d$ and for any $n = 1, 2, \ldots$ there are rationals $\alpha, \beta, \gamma, \delta$ satisfying

$$a < b < \alpha < \beta < \gamma < \delta < c < d$$

$$\delta - \alpha < \frac{1}{n}$$

$$M(\alpha, \beta, \omega) = M(\gamma, \delta, \omega) = M(a, d, \omega).$$

The set of ω such that $B(\cdot, \omega)$ has a nonstrict local maximum is therefore measurable. ★

(54) Lemma. *Let f be a continuous, real-valued function on $[0, \infty)$, such that each local maximum of f is strict. Then f has only countably many local maxima.*

PROOF. Let M_n be the set of x such that

$$f(x) > f(y) \text{ for all } y \text{ with } 0 < |y - x| < 1/n.$$

Each point of M_n is isolated, so M_n is countable. But $\bigcup_n M_n$ is the set of local maxima of f. ★

SUMMARY. The set of ω satisfying the following conditions is measurable and has probability 1:

$$B(\cdot, \omega) \quad \text{is monotone in no interval;}$$

$$\text{all local maxima of } B(\cdot, \omega) \text{ are strict.}$$

For these ω, the local maxima automatically constitute a countable dense set.

Remember

$$m(\varepsilon, \omega) = \min_{0 \leqq t \leqq \varepsilon} B(t, \omega) \quad \text{and} \quad M(\varepsilon, \omega) = \max_{0 \leqq t \leqq \varepsilon} B(t, \omega).$$

(55) Lemma. *For almost all ω: for all $\varepsilon > 0$,*

$$m(\varepsilon, \omega) < 0 < M(\varepsilon, \omega).$$

PROOF. Let $t_1 > t_2 > \cdots > t_n \to 0$. Let $A_n = \{B(t_n) > 0\}$. Then $P\{A_n\} = \frac{1}{2}$. By Fatou, $P\{\limsup A_n\} \geqq \frac{1}{2}$. But $\{\limsup A_n\}$ is in the tail σ-field of the variables

$$\{B(t_n) - B(t_{n+1}) : n = 1, 2, \ldots\}.$$

The Kolmogorov 0–1 law forces $P\{\limsup A_n\} = 1$. But

$$\{\limsup A_n\} \subset \{M(\varepsilon, \cdot) > 0 \text{ for all } \varepsilon > 0\}.$$

Symmetry (11a) handles m. ★

Level sets

Let $S_y(\omega) = \{t : 0 \leqq t < \infty \text{ and } B(t, \omega) = y\}$.

(56) Lemma. *$S_0(\omega)$ has 0 as a limit point, for almost all ω.*

PROOF. $S_0(\omega)$ has 0 as a limit point iff for any rational $b > 0$, there is a rational a with $0 < a < b$ and

$$\inf \{|B(r)| : a \leqq r \leqq b \text{ and } r \text{ is rational}\} = 0.$$

This set of ω is measurable, and contains all ω with the property of (55). ★

(57) Theorem. *For almost all* ω, *the set* $S_0(\omega)$ *is*

(a) *closed*
(b) *unbounded*
(c) *of Lebesgue measure* 0
(d) *dense in itself*
(e) *included in the closure of* $\{t: B(t, \omega) > 0\}$
(f) *included in the closure of* $\{t: B(t, \omega) < 0\}$.

PROOF. The continuity of $B(\cdot, \omega)$ implies (a), for all ω. The continuity and (18) imply (b). Fubini implies (c), because

$$\mathscr{P}\{B(t) = 0\} = 0.$$

I will argue (d) from (56) and the strong Markov property (21). For rational r, let τ_r be the least $t \geqq r$ with $B(t) = 0$: rely on (18). Plainly, τ_r is Markov. Let A_r be the set of ω such that $\tau_r(\omega)$ is a limit point of $S_0(\omega)$. As (56) and strong Markov (21) imply, $\mathscr{P}\{A_r\} = 1$. But $\omega \in \bigcap_r A_r$ iff $S_0(\omega)$ is dense in itself: for $\omega \in \bigcap_r A_r$ iff each point of $S_0(\omega)$ isolated from the left is a limit point from the right. The argument for (d), coupled with (55), shows that the left endpoints of $S_0(\omega)$ are in the closure of $\{t: B(t, \omega) > 0\}$. But $S_0(\omega)$ is perfect and nowhere dense, by (a, d, c). So the left endpoints of $S_0(\omega)$ are dense in $S_0(\omega)$. This proves (e), and (f) is symmetric. ★

The argument for (d) was suggested by Roger Baker. A set is *perfect* iff it is closed and dense in itself.

In view of (42), the set

$$T_0(\omega) = \{t: B(t, \omega) > B(s, \omega) \text{ for all } s < t\}$$

has the same properties as $S_0(\omega)$.

(58) Theorem. *For each* y, *for almost all* ω, *the set* $S_y(\omega)$ *is closed, unbounded, of Lebesgue measure* 0, *and dense in itself.*

PROOF. Let τ be the least t with $B(t) = y$. Clearly, τ is Markov. Use (57) and the strong Markov property (21). ★

(59) Corollary. *For almost all* ω: *for Lebesgue almost all* y, *the set* $S_y(\omega)$ *is perfect.*

PROOF. This follows from (58) and Fubini, after I check that

$$A = \{(y, \omega): S_y(\omega) \text{ is perfect}\}$$

is product measurable. For rational a, b with $0 \leqq a < b$, and $n = 1, 2, \ldots,$ and $m = 0, \ldots, n - 1$, let

$$F_{a,b}(y, \omega) = \inf\{|B(r, \omega) - y| : r \text{ is rational and } a \leqq r \leqq b\}$$

and let $G_{a,b}^{m,n}(y, \omega)$ be the inf of $\{|B(r, \omega) - y|$ for rational r with

$$a + \frac{m}{n}(b - a) \leqq r \leqq a + \frac{m+1}{n}(b - a).$$

Then A is the set of pairs (y, ω) such that for all rational a, b with $0 \leqq a < b$, either $F_{a,b}(y, \omega) > 0$ or for all large enough n, there are three or more $m = 0, \ldots, n - 1$ with $G_{a,b}^{m,n}(y, \omega) = 0$. ★

NOTE. By (51) and (52), for almost all ω, there are $y = y(\omega)$ such that $S_y(\omega)$ has isolated points.

Differentiation

For a real-valued function f on $[0, \infty)$, let

$$D^*f \cdot x = \limsup_{h \downarrow 0} \frac{f(x + h) - f(x)}{h}$$

and

$$D_*f \cdot x = \liminf_{h \downarrow 0} \frac{f(x + h) - f(x)}{h},$$

the *upper and lower right Dini derivates*. If f is continuous, then h can be confined to the rationals without affecting D^*f or D_*f.

(60) Theorem. *Let G be the set of ω such that: for all t,*

$$\text{either} \quad D^*B(t, \omega) = \infty \quad \text{or} \quad D_*B(t, \omega) = -\infty.$$

Then G has inner probability 1. *Namely, G includes an $\mathscr{F}$-set of probability* 1.

I do not know whether G is in $\mathscr{F}$.

PROOF. Let G_1 be the set of ω such that for all $t \in [0, 1]$,

$$\text{either} \quad D^*B(t, \omega) = \infty \quad \text{or} \quad D_*B(t, \omega) = -\infty.$$

It is enough (11b) to prove that G_1 has inner probability 1. Let $A(j, k)$ be the set of ω such that for some t in $[0, 1]$,

$$|B(t + h, \omega) - B(t, \omega)| \leqq jh \quad \text{for all } h \text{ in } [0, 1/k].$$

Now $\bigcup_{j=1}^{\infty} \bigcup_{k=1}^{\infty} A(j, k)$ is the set of ω such that for some t in $[0, 1]$,

$$-\infty < D_*B(t, \omega) \leqq D^*B(t, \omega) < \infty.$$

The problem is to exhibit a measurable $C \supset A(j, k)$ with $\mathscr{P}\{C\} = 0$. Of course, C is allowed to depend on j and k.

Let $C(i, n)$ be the set of ω such that

$$\left| B\left(\frac{i+1}{n}, \omega\right) - B\left(\frac{i}{n}, \omega\right) \right| \leqq \frac{3j}{n}$$

and

$$\left| B\left(\frac{i+2}{n}, \omega\right) - B\left(\frac{i+1}{n}, \omega\right) \right| \leqq \frac{5j}{n}$$

and

$$\left| B\left(\frac{i+3}{n}, \omega\right) - B\left(\frac{i+2}{n}, \omega\right) \right| \leqq \frac{7j}{n}.$$

If $n \geqq 4k$, then $A(j, k) \subset \bigcup_{i=1}^{n} C(i, n)$. Indeed, suppose $t \in [0, 1]$ and

$$|B(t + h, \omega) - B(t, \omega)| \leqq jh \quad \text{for all } h \text{ in } [0, 1/k].$$

Choose i so $(i - 1)/n \leqq t \leqq i/n$. I say $\omega \in C(i, n)$. Indeed, for $v = 0, 1, 2$:

$$\begin{aligned} &\left| B\left(\frac{i+v+1}{n}, \omega\right) - B\left(\frac{i+v}{n}, \omega\right) \right| \\ &\qquad \leqq \left| B\left(\frac{i+v+1}{n}, \omega\right) - B(t, \omega) \right| + \left| B\left(\frac{i+v}{n}, \omega\right) - B(t, \omega) \right| \\ &\qquad \leqq j\frac{2v+3}{n}; \end{aligned}$$

because $\dfrac{i+\mu}{n} - t \leqq \dfrac{\mu+1}{n} \leqq \dfrac{1}{k}$ for $\mu = v$ or $v + 1$.

The probability that an $N(0, 1)$ random variable does not exceed ε in absolute value is at most ε, by (4b). Consequently,

$$\mathscr{P}\{C(i, n)\} \leqq \frac{3 \cdot 5 \cdot 7 \cdot j^3}{n^{3/2}},$$

and $n\,\mathscr{P}\{C(i, n)\} \to 0$. Therefore,

$$C = \bigcap_{n=4k}^{\infty} \bigcup_{i=1}^{n} C(i, n)$$

works. ★

The proof is from (Dvoretzky, Erdos, Kakutani, 1960). This paper is well worth study, although Sections 3–4 must be taken with several grains of salt.

Variation of order 2

Let $S = \{s_0, s_1, \ldots, s_n\}$ be a subset of [0, 1], with $0 \leqq s_0 \leqq \cdots \leqq s_n \leqq 1$. Let

$$\delta S = \max\{|s_{j+1} - s_j| : j = 0, \ldots, n-1\}.$$

Let f be a real-valued function on [0, 1]. Let

$$S^k f = \Sigma_{j=0}^{n-1} |f(s_{j+1}) - f(s_j)|^k.$$

A *partition sequence* Π of [0, 1] is a sequence $\Pi_0 = \{0, 1\} \subset \Pi_1 \subset \Pi_2 \subset \cdots$ of finite subsets of [0, 1]. Let

$$\delta\Pi = \lim_{n\to\infty} \delta\Pi_n \quad \text{and} \quad \Pi^k f = \lim_{n\to\infty} \Pi_n^k f.$$

Call $\Pi^k f$ the *variation of order* k *of* f *with respect to* Π provided the limit exists. For $k = 1$, the sequence $\Pi_n^k f$ is nondecreasing with n; if $\delta\Pi = 0$, the limit does not depend on Π, and is the usual variation of f. In general, $\Pi^k f$ does not exist, and if it does, it depends on Π; for an illustration, see (70).

(61) Theorem. *Let* Π *be a partition sequence of* [0, 1]. *Then* $\Pi^2 B(\cdot, \omega)$ *exists and is finite for almost all* ω. *If* $\delta\Pi = 0$, *then* $\Pi^2 B(\cdot, \omega) = 1$ *for almost all* ω.

For an interesting generalization, see (Cogburn and Tucker, 1961).

Before proving (61), I want to introduce tree diagrams and addition rules. A *tree diagram* T consists of a finite sequence of finite sets $F_0, F_1, \ldots, F_N$, and a linking relation $\to$, with the properties (a, b).

(a) If $n = 0, \ldots, N-1$ and $j \in F_{n+1}$, then $j \to i$ for exactly one $i \in F_n$.

(b) If $n = 0, \ldots, N-1$ and $i \in F_n$, then $j \to i$ for at least one $j \in F_{n+1}$. Typically, there are several $j \to i$.

This is illustrated in Figure 6.

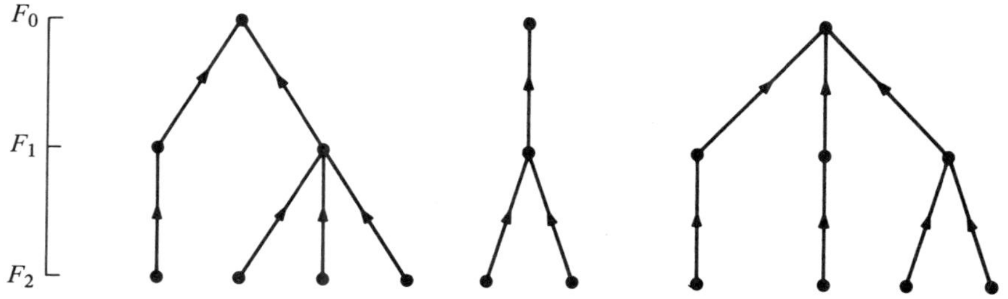

Figure 6.

The linking relation $\to$ defines the chain relation $\Rightarrow$ as follows: $j \Rightarrow i$ iff either $j = i$, or $j \in F_n$ and $i \in F_m$ with $0 \leqq m < n \leqq N$, and there are

$$i_{m+1} \in F_{m+1}, \ldots, i_{n-1} \in F_{n-1},$$

such that

$$j \to i_{n-1} \to \cdots \to i_{m+1} \to i.$$

The chain relation has the following three properties.

(a) If $0 \leqq m < n \leqq N$ and $j \in F_n$, there is exactly one $i \in F_m$ with $j \Rightarrow i$.

(b) If $0 \leqq m < n \leqq N$ and $i \in F_m$, there is at least one $j \in F_n$ with $j \Rightarrow i$.

(c) $k \Rightarrow j$ and $j \Rightarrow i$ imply $k \Rightarrow i$.

For (62–64), let $(F_0, \ldots, F_N; \to)$ be a tree diagram. Suppose f is a function on F_N. Then f defines an *addition rule* f^+ on $F = F_0 \cup \cdots \cup F_N$ as follows:

$$f^+(i) = \Sigma_j \{f(j) : j \in F_N \text{ and } j \Rightarrow i\}.$$

If $0 \leqq m < n \leqq N$ and $i \in F_m$, then

$$f^+(i) = \Sigma_j \{f^+(j) : j \in F_n \text{ and } j \Rightarrow i\}.$$

This characterizes addition rules.

ILLUSTRATION. Let $N = 1$ and $F_0 = \{a\}$ and $F_1 = \{b, c\}$ with $b \to a$ and $c \to a$. Then

$$f^+(a) = f(b) + f(c)$$
$$f^+(b) = f(b)$$
$$f^+(c) = f(c).$$

For (62), suppose $b \in F_1$. Let f be a function on F_N. Define

$$\begin{aligned} g(j) &= -f(j) \quad \text{for } j \in F_N \text{ and } j \Rightarrow b \\ &= f(j) \quad \text{for } j \in F_N \text{ and } j \not\Rightarrow b. \end{aligned}$$

Then g is also a function on F_N, and determines its own addition rule g^+.

(62) Lemma. *Suppose $i \in F_1 \cup \cdots \cup F_N$. Then $g^+(i)$ is $-f^+(i)$ or $f^+(i)$, according as $i \Rightarrow b$ or $i \not\Rightarrow b$.*

PROOF. Suppose $i \Rightarrow b$. Then $j \in F_N$ and $j \Rightarrow i$ imply $j \Rightarrow b$, so $g(j) = -f(j)$ and

$$\begin{aligned} g^+(i) &= \Sigma_j \{g(j) : j \in F_N \text{ and } j \Rightarrow i\} \\ &= \Sigma_j \{-f(j) : j \in F_N \text{ and } j \Rightarrow i\} \\ &= -f^+(i). \end{aligned}$$

The argument for $i \not\Rightarrow b$ is similar, because $j \in F_N$ and $j \Rightarrow i$ imply $j \not\Rightarrow b$. ★

For (63–64), suppose $\{Y(j): j \in F_N\}$ are independent and have symmetric distributions, on $(\Omega, \mathscr{F}, \mathscr{P})$. *Symmetry* means $-Y(j)$ is distributed like $Y(j)$. Define a random addition rule Y^+ on $F = F_0 \cup \cdots \cup F_N$:

$$Y^+(i) = \Sigma_j \{Y(j): j \in F_N \text{ and } j \Rightarrow i\}.$$

Let

$$S_n = \Sigma_j \{Y^+(j)^2 : j \in F_n\}.$$

Suppose $b \neq c$ are in F_1.

(63) Lemma. $E\{Y^+(b)Y^+(c)|S_1, S_2, \ldots, S_N\} = 0$ *almost surely.*

PROOF. Let

$$\begin{aligned} Z(j) &= -Y(j) \quad \text{for } j \in F_N \text{ and } j \Rightarrow b \\ &= Y(j) \qquad \text{for } j \in F_N \text{ and } j \not\Rightarrow b. \end{aligned}$$

Then Z determines its own random addition rule Z^+. Let

$$T_n = \Sigma_i \{Z^+(i)^2 : i \in F_n\}.$$

Now Y is distributed like Z, so Y^+ is distributed like Z^+, and

$$Y^+(b), Y^+(c), S_1, \ldots, S_N$$

is distributed like

$$Z^+(b), Z^+(c), T_1, \ldots, T_N.$$

By (62), for $i \in F_1 \cup \cdots \cup F_N$,

$$\begin{aligned} Z^+(i) &= -Y^+(i) \quad \text{when } i \Rightarrow b \\ &= Y^+(i) \qquad \text{when } i \not\Rightarrow b. \end{aligned}$$

So,

$$Z^+(b) = -Y^+(b) \quad \text{and} \quad Z^+(c) = Y^+(c) \quad \text{and}$$
$$T_n = S_n \quad \text{for} \quad n = 1, \ldots, N.$$

Therefore,

$$Y^+(b), Y^+(c), S_1, \ldots, S_N$$

is distributed like

$$-Y^+(b), Y^+(c), S_1, \ldots, S_N.$$

And with probability 1,

$$E\{Y^+(b)Y^+(c)|S_1, \ldots, S_N\} = E\{-Y^+(b)Y^+(c)|S_1, \ldots, S_N\}. \qquad \bigstar$$

(64) Lemma. $E\{S_0|S_1, \ldots, S_N\} = S_1$ *almost surely.*

PROOF. For $a \in F_0$,

$$Y^+(a) = \Sigma_j \{Y^+(j) : j \in F_1 \text{ and } j \to a\}.$$

So

$$Y^+(a)^2 = U(a) + 2W(a),$$

where

$$U(a) = \Sigma_j \{Y^+(j)^2 : j \in F_1 \text{ and } j \to a\},$$

$$W(a) = \Sigma_{b,c} \{Y^+(b)Y^+(c) : b, c \in F_1 \text{ and } b, c \to a \text{ but } b \neq c\}.$$

Now

$$\Sigma_a \{U(a) : a \in F_0\} = S_1.$$

And

$$E\{W(a)|S_1, \ldots, S_N\} = 0 \quad \text{almost surely}$$

by (63). But $S_0 = \Sigma_a \{Y^+(a)^2 : a \in F_0\}$. ★

PROOF OF (61). Remember that $\{\Pi_n\}$ is a partition sequence. Let

$$V_n(\omega) = \Pi_n^2 B(\cdot, \omega),$$

so $V_0 = B(1)^2$. The main point is proving

(65) $$\ldots, V_2, V_1, V_0$$

is a martingale. That is,

(66) $$E\{V_k|V_{k+1}, V_{k+2}, \ldots\} = V_{k+1} \quad \text{almost surely}$$

for all $k = 0, 1, \ldots$. By forward martingales (3.34), this reduces to

(67) $$E\{V_k|V_{k+1}, V_{k+2}, \ldots, V_{k+N}\} = V_{k+1} \quad \text{almost surely,}$$

for all k and $N = 0, 1, \ldots$.

I will now argue (67). Fix k and N. For $n = 0, \ldots, N$, let F_n be the set of intervals into which Π_{k+n} partitions $[0, 1]$. For definiteness, make all these intervals closed. For $i \in F_n$ and $j \in F_{n+1}$, let $j \to i$ iff $j \subset i$. Check that $(F_0, \ldots, F_N; \to)$ is a tree diagram. For $j \in F_N$, let $Y(j)$ be the increment of B over j: if $j = [s, t]$,

$$Y(j) = B(t) - B(s).$$

These $Y(j)$ are independent and symmetric. If $j \in F_0 \cup \cdots \cup F_N$, then

$Y^+(j)$ is the increment of B over j. Let

$$S_n = \Sigma_j \{Y^+(j)^2 : j \in F_n\}.$$

Then $S_n = V_{k+n}$, and (67) is a special case of (64).

Granting (67), backward martingales (3.34) show that V_k converges almost surely to a limit $V \in L^2$, and

$$E\{(V - V_k)^2\} \to 0.$$

This proves the first part of the theorem. For the second, enumerate Π_k as

$$0 = t_{k,0} < t_{k,1} < \cdots < t_{k,n(k)} = 1.$$

Then

$$V_k = \Sigma_{m=1}^{n(k)} Y_{k,m}^2,$$

where

$$Y_{k,m} = B(t_{k,m}) - B(t_{k,m-1}).$$

Abbreviate $v = \operatorname{var} B(1)^2$, which I am too lazy to compute. Now

$$\begin{aligned} E(V_k) &= \Sigma_{m=1}^{n(k)} E(Y_{k,m}^2) \\ &= \Sigma_{m=1}^{n(k)} (t_{k,m} - t_{k,m-1}) \\ &= 1. \end{aligned}$$

And

$$\begin{aligned} \operatorname{var}(V_k) &= \Sigma_{m=1}^{n(k)} \operatorname{var}(Y_{k,m}) \\ &= \Sigma_{m=1}^{n(k)} v(t_{k,m} - t_{k,m-1})^2 \\ &\leqq v\delta_k, \end{aligned}$$

where

$$\delta_k = \max_m (t_{k,m} - t_{k,m-1}).$$

If $\delta_k \to 0$, then $V_k \to 1$ in L^2, so $V = 1$ almost surely. ★

This proof was suggested by Doob (1953, p. 395).

To appreciate (61), consider the following facts about a continuous function f on [0, 1].

(68) Lemma. *If Π is a partition sequence with $\delta\Pi = 0$ and f is of bounded variation, then $\Pi^2 f = 0$.*

PROOF. Let $\operatorname{osc}(f, \delta) = \max\{|f(x) - f(y)| : |x - y| \leqq \delta\}$. Check

$$\Pi_n^2 f \leqq (\Pi_n f) \cdot \operatorname{osc}(f, \delta\Pi_n) \to 0.$$ ★

(69) Lemma. *For $\varepsilon > 0$, there is a finite set $S = S(\varepsilon, f)$ with*

$$\{0, 1\} \subset S \subset [0, 1] \text{ and } S^2 f < \varepsilon.$$

PROOF. *Case 1*: $f(1) > f(0)$. See Figure 7.

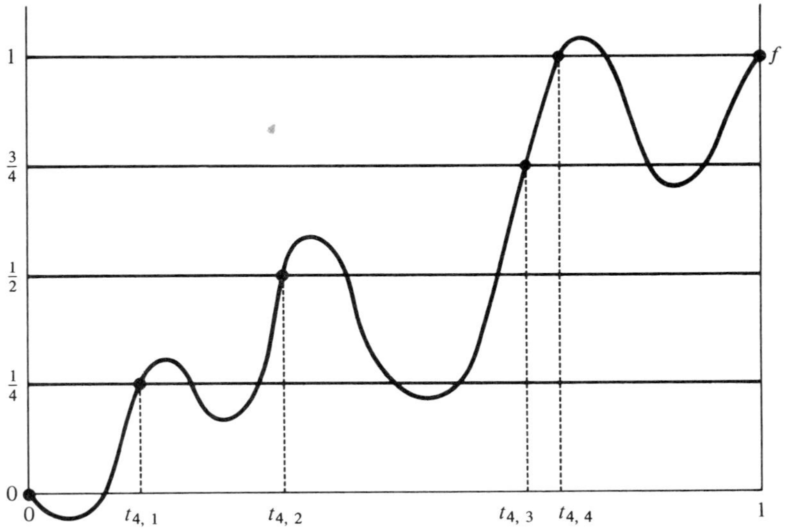

Figure 7.

It is harmless to suppose $f(0) = 0$ and $f(1) = 1$. Let $t_{n,j}$ be the least t with $f(t) = j/n$, for $j = 0, \ldots, n$. Let $S_n = \{t_{n,0}, \ldots, t_{n,n}\}$ if $t_{n,n} = 1$, and let $S_n = \{t_{n,0}, \ldots, t_{n,n}, 1\}$ otherwise. Clearly, $S_n^2 f = 1/n$.

Case 2: $f(1) < f(0)$. Use Case 1 on $g(x) = f(1 - x)$.

Case 3: $f(0) = f(1)$. Let $S = \{0, 1\}$. ★

(70) Proposition. *There is a partition sequence Π with $\delta\Pi = 0$ and $\Pi^2 f = 0$.*

PROOF. Fix a sequence $\varepsilon_n \downarrow 0$. Let $\Pi_0 = \{0, 1\}$. Suppose Π_n is defined. Let F_{n+1} be a finite set with $\Pi_n \subset F_{n+1} \subset [0, 1]$ and $\delta F_{n+1} \leqq \varepsilon_{n+1}$. Suppose F_{n+1} has $m + 1$ points, dividing $[0, 1]$ into m intervals. Using (69), construct a finite set Π_{n+1} with $F_{n+1} \subset \Pi_{n+1} \subset [0, 1]$, such that for each pair (a, b) of successive points of F_{n+1},

$$(\Pi_{n+1} \cap [a, b])^2 f \leqq \varepsilon_{n+1}/m.$$

Thus, $\delta\Pi_{n+1} \leqq \delta F_{n+1} \leqq \varepsilon_{n+1}$ and $\Pi_{n+1}^2 f \leqq \varepsilon_{n+1}$. ★

I learned this argument from David Blackwell.

Let $e(\Pi)$ be the set of ω such that

$$\Pi^2 B(\cdot, \omega) \neq 1.$$

Then $\mathscr{P}\{e(\Pi)\} = 0$ for each Π by (61), but $\bigcup_\Pi e(\Pi) = \Omega$ by (70).

According to (Lévy, 1965, p. 190): for almost all ω, there is a partition sequence $\Pi(\omega)$ of $[0, 1]$, with $\delta\Pi(\omega) = 0$ and $\Pi(\omega)^2 B(\cdot, \omega) = \infty$. Here is a sketch of the argument Dick Drogin and I found. For positive integer k, let

$$p(k) = \tfrac{1}{2}\mathscr{P}\{B(1)^2 \geqq 2k\} > 0.$$

Say an interval $[s, t]$ has *ω-weight k* iff

$$[B(t, \omega) - B(s, \omega)]^2 \geqq 2k(t - s).$$

The probability that

$$\left[\frac{i-1}{n}, \frac{i}{n}\right] \text{ has weight } k \text{ for at least } p(k)n \text{ indices } i = 1, \ldots, n$$

tends to 1 as $n \to \infty$. Consequently, the event that

$\left[\frac{i-1}{n}, \frac{i}{n}\right]$ has weight k for at least $p(k)n$ indices $i = 1, \ldots, n$, for some positive integer n

has probability 1. For positive integer M, n, k, let $G(M, n, k)$ be the event that

$\left[\frac{m}{M} + \frac{i-1}{Mn}, \frac{m}{M} + \frac{i}{Mn}\right]$ has weight k for at least $p(k)n$ indices $i = 1, \ldots, n$, for all $m = 0, \ldots, M - 1$.

Let

$$G = \bigcap_{M=1}^\infty \bigcap_{k=1}^\infty \bigcup_{n=1}^\infty G(M, n, k).$$

Then $\mathscr{P}(G) = 1$.

Fix $\omega \in G$. Fix positive integers M and k. I will construct a finite subset F of the rationals in $[0, 1]$, which includes $\{m/M : m = 0, \ldots, M\}$, while the sum of the lengths of the F-intervals of ω-weight k is at least $\frac{1}{2}$. So $F^2 B(\cdot, \omega) \geqq k$. A rational interval of *order* n is an interval $[(i-1)/n, i/n]$. To begin with, there is a positive integer n_1 with $\omega \in G(M, n_1, k)$. Let I_1 be the set of rational intervals of order Mn_1 and ω-weight k. Define $U_1 = \bigcup\{I : I \in I_1\}$. So

$$\text{Lebesgue }\{[0, 1] \setminus U_1\} \leqq 1 - p(k).$$

There is a positive integer n_2 with $\omega \in G(Mn_1, n_2, k)$. Let I_2 be the set of rational intervals of order Mn_1n_2 and ω-weight k, which are not subsets of U_1. Let $U_2 = \bigcup\{I : I \in I_1 \cup I_2\}$. So

$$\text{Lebesgue }\{[0, 1] \setminus U_2\} \leqq [1 - p(k)]^2.$$

Keep going N times, where

$$[1 - p(k)]^N \leqq \tfrac{1}{2}.$$

Let F consist of the endpoints of the intervals in $I_1 \cup \cdots \cup I_N$, and $\{m/M : m = 0, \ldots, M\}$. ★

The next result improves (48).

(71) Theorem. *For almost all ω, the function $B(\cdot, \omega)$ is of bounded variation in no interval.*

FIRST PROOF. Use (61) and (68). ★

SECOND PROOF. Use (60) and Lebesgue's theorem (3.60). ★

THIRD PROOF. Using (48) and (59), confine ω to the set of probability 1 where

(a) $\qquad B(\cdot, \omega)$ is monotone on no interval

and

(b) $\qquad S_y(\omega)$ is perfect for almost all y.

Suppose $s < t$. Then $B(\cdot, \omega)$ assumes an interval of values on $[s, t]$, by (a). With this and (b) for moral support, the set of y for which

$$S_y(\omega) \cap [s, t] \quad \text{is nonempty and perfect}$$

has positive Lebesgue measure. But a nonempty perfect set has the power of the continuum, by Cantor's theorem (Hausdorff, 1957, p. 156). Now use Banach's theorem (3.63). ★

5. STRASSEN'S LAW OF THE ITERATED LOGARITHM

Preliminaries

Remember that $C[0, 1]$ is the set of continuous functions on $[0, 1]$. For $f \in C[0, 1]$, let

$$\|f\| = \max \{|f(t)| : 0 \leqq t \leqq 1\}.$$

Give $C[0, 1]$ the metric d:

$$d(f, g) = \|f - g\|.$$

Then $C[0, 1]$ is complete and separable. Let Σ be the smallest σ-field of subsets of $C[0, 1]$ which includes the open sets. Then Σ is Borel. Remember that

$$\xi(t, f) = f(t) \quad \text{for } f \in C[0, 1] \text{ and } t \in [0, 1].$$

Let Σ^* be the smallest σ-field relative to which all $\xi(t)$ are measurable.

(72) Lemma. $\Sigma = \Sigma^*$.

PROOF. Clearly, $\xi(t)$ is continuous, so Σ-measurable. That is, $\Sigma^* \subset \Sigma$. The converse is subtler. For any $g \in C[0, 1]$,

$$\|f - g\| = \sup\{|f(r) - g(r)| : 0 \leqq r \leqq 1 \text{ and } r \text{ is rational}\};$$

so $f \to \|f - g\|$ is Σ^*-measurable. The spheres

$$S(g, r) = \{f : f \in C[0, 1] \text{ and } \|f - g\| < r\}$$

are in Σ^*. An open set is a countable union of spheres, and is in Σ^*. That is, $\Sigma \subset \Sigma^*$. ★

To state (73), suppose $Y(t, \cdot)$ is an $\mathscr{F}$-measurable function on Ω for each $t \in [0, 1]$. Suppose $Y(\cdot, \omega)$ is a continuous function on $[0, 1]$ for each $\omega \in \Omega$. Then Y can be visualized as a mapping from Ω to $C[0, 1]$, which assigns to $\omega \in \Omega$ the function $Y(\cdot, \omega) \in C[0, 1]$. If H is a subset of $C[0, 1]$, then $Y^{-1}H$ is a subset of Ω:

$$Y^{-1}H = \{\omega : \omega \in \Omega \text{ and } Y(\cdot, \omega) \in H\}.$$

If $g \in C[0, 1]$, then $\|Y - g\|$ is a function on Ω, whose value at ω is $\|Y(\cdot, \omega) - g\|$.

(73) Lemma. **(a)** *Y is a measurable mapping from $(\Omega, \mathscr{F})$ to*

$$(C[0, 1], \Sigma).$$

(b) *If $H \in \Sigma$, then $Y^{-1}H \in \mathscr{F}$.*

(c) *$\|Y - g\|$ is a measurable function on $(\Omega, \mathscr{F})$, for $g \in C[0, 1]$.*

PROOF. *Claim (a).* By (72), you only have to check that $\xi(t) \circ Y$ is measurable for each t. But I gave you that.

Claim (b) follows from (a).

Claim (c) follows from (b). ★

Compactness

A subset H of $C[0, 1]$ is *compact* iff every sequence in H has a subsequence which converges to an element of H. A subset H of $C[0, 1]$ is *relatively compact* iff every sequence in H has a subsequence which converges, perhaps to a limit outside H. To say this another way, H is relatively compact iff the closure of H is compact: the closure of H being the smallest closed superset of H. Let $H = \{f_n : n \in N\}$ be an indexed subset of $C[0, 1]$. The *limit set* L of H

is the set of $g \in C[0, 1]$ such that: for any positive ε, there are arbitrarily large indices n with $\|f_n - g\| < \varepsilon$. The limit set L of H is necessarily closed, but it may be a little smaller than the closure $\bar{H}$ of H, and may depend on the indexing. Namely: $L \setminus H = \bar{H} \setminus H$, and $H \subset \bar{H}$; but L does not necessarily include H. Indeed, fix $g \in H$. Let N_g be the set of $n \in N$ with $f_n = g$. Then $g \in L$ iff either N_g contains arbitrarily large indices, or g is in the closure of $\{f_n : n \in N \setminus N_g\}$.

To state (74–75), introduce the space $C[0, 1]^\infty$ of sequences

$$\phi = (\phi_0, \phi_1, \ldots)$$

of elements of $C[0, 1]$. Give $C[0, 1]^\infty$ the product σ-field Σ^∞. For (74), let $Y_0, Y_1, \ldots$ be measurable maps from $(\Omega, \mathscr{F})$ to $(C[0, 1], \Sigma)$. Let

$$Y = (Y_0, Y_1, \ldots),$$

a mapping from Ω to $C[0, 1]^\infty$.

(74) Lemma. *Y is measurable.*

PROOF. Easy. ★

For (75), let H be a compact subset of $C[0, 1]$. Let H^* be the set of $\phi \in C[0, 1]^\infty$ such that

the indexed subset $(\phi_n : n = 0, 1, \ldots)$ of $C[0, 1]$ is relatively compact, with limit set H.

(75) Lemma. *$H^* \in \Sigma^\infty$.*

PROOF. To begin with, H has a countable, dense subset H_0. Define

$$\begin{aligned} d(f, H) &= \inf \{d(f, g) : g \in H\} \\ &= \inf \{d(f, g) : g \in H_0\}. \end{aligned}$$

So, $f \to d(f, H)$ is Σ-measurable. For $\phi \in C[0, 1]^\infty$, let

$$d^*(\phi, H) = \limsup_{n\to\infty} d(\phi_n, H).$$

So, $\phi \to d^*(\phi, H)$ is Σ^∞-measurable. For $\phi \in C[0, 1]^\infty$ and $g \in C[0, 1]$, let

$$d_*(\phi, g) = \liminf_{n\to\infty} d(\phi_n, g).$$

So, $\phi \to d_*(\phi, g)$ is Σ^∞-measurable. Let

$$A = \{\phi : \phi \in C[0, 1]^\infty \text{ and } d^*(\phi, H) = 0 \text{ and } d_*(\phi, g) = 0 \text{ for all } g \in H_0\}.$$

Clearly, $A \in \Sigma^\infty$. I say

(76) $$H^* = A.$$

You check $H^* \subset A$. For the converse, fix $\phi \in A$. Because $d^*(\phi, H) = 0$, there are $f_n \in H$ with

$$\|\phi_n - f_n\| \to 0.$$

Use the compactness of H to find a subsequence n' with $f_{n'} \to f \in H$. Then $\phi_{n'} \to f$. This proves

$$\{\phi_n : n = 0, 1, \ldots\}$$

is relatively compact, with limit set $L(\phi) \subset H$. Now $L(\phi) \supset H_0$, because $d_*(\phi, g) = 0$ for all $g \in H_0$. But limit sets are closed and H_0 is dense in H, so $L(\phi) \supset H$. That is, $\phi \in H^*$. ★

(77) Definition. *Let K be the set of absolutely continuous functions f on* $[0, 1]$, *with* $f(0) = 0$ *and*

$$\int_0^1 f'(t)^2 \, dt \leqq 1.$$

(78) Lemma. *Suppose $f \in K$.*

(a) *If $0 \leqq s \leqq t \leqq 1$, then $|f(t) - f(s)| \leqq (t - s)^{\frac{1}{2}}$.*

(b) *If $0 \leqq t \leqq 1$, then $|f(t)| \leqq t^{\frac{1}{2}}$.*

(c) $\|f\| \leqq 1$.

(d) *K is compact.*

PROOF. *Claim (a)* follows from the Schwarz inequality:

$$f(t) - f(s) = \int_s^t f'(u) \, du;$$

$$(f(t) - f(s))^2 \leqq \int_s^t 1 \, du \cdot \int_s^t f'(u)^2 \, du \leqq t - s.$$

Claim (b) follows from (a).

Claim (c) follows from (b).

Claim (d). Let $f_n \in K$. The unit ball of L^2 is weak* compact (Royden, 1963, p. 173 for the theorem and Chapter 6 for the background); so there is a subsequence n^* and a measurable function g' on $[0, 1]$ such that the sequence f'_{n^*} weak* converges to g' and

$$\int_0^1 g'(t)^2 \, dt \leqq 1.$$

Let

$$g(t) = \int_0^t g'(s) \, ds,$$

so $g \in K$. Now

$$f_{n*}(t) = \int_0^t f'_{n*}(s)\, ds$$

is the inner product of f'_{n*} with the indicator function of the interval $[0, t]$, and converges to $g(t)$. That is, f'_{n*} converges to g pointwise. But $\{f_n, g\}$ is equicontinuous by (b), so the convergence is uniform. ★

(79) Lemma. *Let $a > 1$ and let $f \in C[0, 1]$ with $a^{-1}f \in K$. Then $\|f\| \leqq a$ and $\|f - a^{-1}f\| \leqq a - 1$.*

PROOF. *First inequality.*

$$\|f\| = \|aa^{-1}f\| = a\|a^{-1}f\| \leqq a, \quad \text{by (78c).}$$

Second inequality.

$$\|f - a^{-1}f\| = \|(a - 1)a^{-1}f\| = (a - 1)\|a^{-1}f\| \leqq a - 1, \quad \text{by (78c).} \ ★$$

To state (80), let f be absolutely continuous on $[0, 1]$, and let

$$0 = t_0 < t_1 < \cdots < t_n = 1.$$

Define $g \in C[0, 1]$ by the requirements:

$$g(t_m) = f(t_m) \qquad \text{for } m = 0, \ldots, n$$

$$g \quad \text{is linear on} \quad [t_{m-1}, t_m] \qquad \text{for } m = 1, \ldots, n.$$

(80) Lemma. $\int_0^1 g'(t)^2\, dt \leqq \int_0^1 f'(t)^2\, dt$. *Equality holds iff $f = g$.*

PROOF. Let $0 \leqq s < t \leqq 1$. As in (78a), Schwarz makes

$$\frac{[f(t) - f(s)]^2}{t - s} \leqq \int_s^t f'(u)^2\, du.$$

Equality holds iff f is linear on $[s, t]$. Put $s = t_{m-1}$ and $t = t_m$; sum over $m = 1, \ldots, n$. The sum of the left sides is $\int_0^1 g'(u)^2\, du$. The sum of the right sides is $\int_0^1 f'(u)^2\, du$. ★

The main result

Remember that B is normalized Brownian motion on the triple $(\Omega, \mathscr{F}, \mathscr{P})$. Define

$$Z_n(t) = (2n \log \log n)^{-\frac{1}{2}} B(nt) \quad \text{for } 0 \leqq t \leqq 1.$$

Lemma (73a) makes it legitimate to view Z_n as a measurable map from $(\Omega, \mathscr{F})$ to $(C[0, 1], \Sigma)$. Strassen's (1964) iterated logarithm is

(81) Theorem. *For almost all ω, the indexed subset*

$$\{Z_n(\cdot, \omega): n = 3, 4, \ldots\}$$

of $C[0, 1]$ *is relatively compact, with limit set* K.

The set K was defined in (77). Your scruples about measurability must yield to the sequence (74), (75), (78d).

A heuristic argument

Let $r > 1$. Classical proofs of the iterated logarithm suggest:

(a) Z_n does not change much as n increases from r^j to r^{j+1}, for small r;

(b) the Z_n are nearly independent as n moves through $1, r, r^2, \ldots$, for large r.

From (a), the limiting behavior of $\{Z_n: n = 1, 2, 3, \ldots\}$ nearly coincides with that of $\{Z_n: n = 1, r, r^2, \ldots\}$. From (b), the limiting behavior of $\{Z_n: n = 1, r, r^2, \ldots\}$ nearly coincides with that of $\{Z_n^*: n = 1, r, r^2, \ldots\}$, where

$$Z_n^*(t) = (2 \log \log n)^{-\frac{1}{2}} B_n(t) \quad \text{for } 0 \leqq t \leqq 1$$

B_n is normalized Brownian motion

$B_1, B_2, \ldots$ are independent.

Fix $m = 1, 2, \ldots$ and $r > 1$. Let $n = r^j$. Let V_j be the m-vector

$$\left\{Z_n^*\left(\frac{i}{m}\right) - Z_n^*\left(\frac{i-1}{m}\right) : i = 1, \ldots, m\right\}.$$

As Borel–Cantelli implies, the length of V_j exceeds $1 + \varepsilon$ for only finitely many j, with probability 1. So $\{V_j : j = 1, 2, \ldots\}$ is relatively compact, with probability 1. The density of V_j at $(x_1, \ldots, x_m)$ is of the order

$$(1/j)^{m\Sigma_{i=1}^m x_i^2}$$

The sum on j is infinite or finite, according as $\Sigma_{i=1}^m x_i^2 \leqq 1/m$ or $\Sigma_{i=1}^m x_i^2 > 1/m$. Now Borel–Cantelli makes

$$\{(x_1, \ldots, x_m) : \Sigma_{i=1}^m x_i^2 \leqq 1/m\}$$

the limit set of $\{V_j : j = 1, 2, \ldots\}$, with probability 1. This limit set doesn't depend on r, because $\log \log r^j$ really doesn't.

Back to business

The formal proof of (81) breaks up into two parts, as (76) suggests:

(a) to show that Z_n approaches K;

(b) to show that Z_n approximates a countable dense subset of K.

Part (a) of the proof: motivation

The natural impulse is to compute

$$\int_0^1 Z_n'(t)^2\,dt.$$

This doesn't work, because B isn't differentiable (60). The local roughness of B drowns out the global smoothness of Z_n. This suggests trying for

$$\int_0^1 Z_{n,m}'(t)^2\,dt,$$

where $Z_{n,m}$ is locally smooth and converges to Z_n as $m \to \infty$. A system $Z_{n,m}$ that works is defined by the requirements

$$Z_{n,m}(i/m) = Z_n(i/m) \quad \text{for } i = 0, \ldots, m$$

$$Z_{n,m}(\cdot) \quad \text{is linearly interpolated.}$$

To simplify the typesetter's life, I will often write $Z(n)$ for Z_n and $Z(n, m)$ for $Z_{n,m}$.

I will argue *first* that for each m and each $b > 1$, the point $Z(n, m)$ approaches the set K as n increases through the geometric sequence $1, b, b^2, \ldots$. *Second*, for large m and each $b > 1$,

$$\|Z(n, m) - Z(n)\|$$

becomes small—but does not tend to 0—as n increases through $1, b, b^2, \ldots$. *Third*, for b only a little larger than 1,

$$\|Z(n) - Z(b^j)\|$$

is eventually small—but does not tend to 0—as n and j increase with $b^j \leqq n < b^{j+1}$. Consequently:

$Z(b^j, m)$ is close to K

$Z(b^j)$ is close to $Z(b^j, m)$

$Z(n)$ is close to $Z(b^j)$, provided $b^j \leqq n \leqq b^{j+1}$;

and you safely conclude

$Z(n)$ is close to K.

Part (a) of proof: formalities

Remember

$$Z(n)(t,\omega) = Z_n(t,\omega) = (2n \log\log n)^{-\frac{1}{2}} B(nt,\omega)$$

for $0 \leqq t \leqq 1$. Remember that $Z(n,m)(\cdot,\omega) = Z_{n,m}(\cdot,\omega) \in C[0,1]$ is defined by the requirements:

$$Z(n,m)\left(\frac{i}{m},\omega\right) = Z_n\left(\frac{i}{m},\omega\right) \qquad \text{for } i = 0,\ldots,m$$

$$Z(n,m)(\cdot,\omega) \quad \text{is linear on} \quad \left[\frac{i-1}{m},\frac{i}{m}\right] \text{ for } i = 1,\ldots,m.$$

I will think of $Z(n)$ and $Z(n,m)$ as measurable maps from $(\Omega,\mathscr{F})$ to $(C[0,1],\Sigma)$, which is my privilege by (73a).

To state (82), let $a > 1$ and $b > 1$ and $m = 1, 2, \ldots$. Let $G_1(a,b,m)$ be the set of $\omega \in \Omega$ for which there exists a

$$J = J_1(a,b,m,\omega) < \infty$$

such that:

$$j \geqq J \quad \text{implies} \quad a^{-1}Z(b^j,m)(\cdot,\omega) \in K.$$

(82) Lemma. *$G_1(a,b,m) \in \mathscr{F}$ and $\mathscr{P}\{G_1(a,b,m)\} = 1$.*

PROOF. Of course, $a^{-1}f \in K$ iff f is absolutely continuous on $[0,1]$, and $f(0) = 0$, and

$$\int_0^1 f'(t)^2\,dt \leqq a^2.$$

Let

$$V_{n,m} = \int_0^1 Z'_{n,m}(t)^2\,dt.$$

Temporarily, fix n and m. Define

$$Y_i = B\left(\frac{i}{m}n\right) - B\left(\frac{i-1}{m}n\right) \quad \text{for } i = 1,\ldots,m.$$

For $\dfrac{i-1}{m} < t < \dfrac{i}{m}$, an interval of length $\dfrac{1}{m}$,

$$Z'_{n,m}(t) = m\left[Z_n\left(\frac{i}{m}\right) - Z_n\left(\frac{i-1}{m}\right)\right]$$
$$= m(2n\log\log n)^{-\frac{1}{2}}\,Y_i.$$

So,

$$V_{n,m} = m(2n \log\log n)^{-1} \Sigma_{i=1}^{m} Y_i^2.$$

Caution. The index $-\frac{1}{2}$ changed to -1.

Now $Y_1, \ldots, Y_m$ are independent and $N(0, n/m)$. So

$$(2 \log\log n) V_{n,m}$$

has the chi-square distribution with m degrees of freedom (Cramer, 1957, Sec. 18.1). This distribution has a density of the form

$$x \to \text{const. } x^{(m/2)-1} e^{-(x/2)}.$$

To be crude, fix any a' with $1 < a' < a^2$. There is a constant $k < \infty$, depending on a, a', b, and m, but not on n, such that

$$\mathscr{P}\{V_{n,m} > a^2\} < k \exp(-a' \log\log n),$$

where $\exp(x) = e^x$. Along $n = b^j$, this bound sums, being of the order $1/j^{a'}$. Use Borel-Cantelli. ★

To state (83), let $\varepsilon > 0$ and $b > 1$ and $m = 1, 2, \ldots$. Let $G_2(\varepsilon, b, m)$ be the set of $\omega \in \Omega$ for which there is a

$$J = J_2(\varepsilon, b, m, \omega) < \infty$$

such that:

$$j \geqq J \quad \text{implies} \quad \|Z(b^j, m)(\cdot, \omega) - Z(b^j)(\cdot, \omega)\| \leqq \varepsilon.$$

(83) Lemma. *$G_2(\varepsilon, b, m) \in \mathscr{F}$. If $m\varepsilon^2 > 1$, then $\mathscr{P}\{G_2(\varepsilon, b, m)\} = 1$.*

PROOF. Temporarily, fix n and m and ε. For $i = 1, \ldots, m$, let A_i be the event that

$$\max\left\{|Z_{n,m}(t) - Z_n(t)| : \frac{i-1}{m} \leqq t \leqq \frac{i}{m}\right\} > \varepsilon.$$

Thus

$$\{\|Z_{n,m} - Z_n\| > \varepsilon\} = \bigcup_{i=1}^{m} A_i.$$

As (11c) implies,

$$\mathscr{P}\{A_1\} = \cdots = \mathscr{P}\{A_m\}.$$

I say; A_1 is the event that

$$\max\left\{\left|B(s) - \frac{m}{n} s B\left(\frac{n}{m}\right)\right| : 0 \leqq s \leqq \frac{n}{m}\right\} > \varepsilon(2n \log\log n)^{\frac{1}{2}}.$$

To check this, remember

$$Z_n(t) = (2n \log\log n)^{-\frac{1}{2}} B(nt)$$

and for $0 \leqq t \leqq 1/m$,

$$Z_{n,m}(t) = mtZ_n\left(\frac{1}{m}\right)$$

$$= mt(2n \log \log n)^{-\frac{1}{2}} B\left(\frac{n}{m}\right).$$

Put $nt = s$.

Now I say $A_1 \subset C \cup D$, where C is the event that

$$\max\left\{|B(s)|: 0 \leqq s \leqq \frac{n}{m}\right\} > \varepsilon(2n \log \log n)^{\frac{1}{2}}$$

and D is the event that

$$\max\left\{\left|B(s) - B\left(\frac{n}{m}\right)\right| : 0 \leqq s \leqq \frac{n}{m}\right\} > \varepsilon(2n \log \log n)^{\frac{1}{2}}.$$

Indeed, $B(s) - \frac{m}{n} s B\left(\frac{n}{m}\right)$ is a convex combination of $B(s)$ and $B(s) - B\left(\frac{n}{m}\right)$; because $0 \leqq s \leqq \frac{n}{m}$. As (11d) implies, $\mathscr{P}\{C\} = \mathscr{P}\{D\}$. As (16c) shows, this common probability is at most

$$4\,\mathscr{P}\left\{B\left(\frac{n}{m}\right) > \varepsilon(2n \log \log n)^{\frac{1}{2}}\right\} = 4\,\mathscr{P}\{B(1) > \varepsilon(2m \log \log n)^{\frac{1}{2}}\}.$$

Caution. One n changed to m.

By (4a), the last display is at most a constant times $\exp(-m\varepsilon^2 \log \log n)$, where $\exp x = e^x$. So

$$\mathscr{P}\{\|Z_{n,m} - Z_n\| > \varepsilon\} \leqq k \exp\,(-m\varepsilon^2 \log \log n),$$

where $k < \infty$ depends on m and ε, but not on n. Along $n = b^j$, this bound sums, being of the order $1/j^\theta$ with $\theta = m\varepsilon^2$. Borel-Cantelli works again. ★

To state (84), let $\delta > 0$ and $b > 1$. Let

$$b(n, j) = (b^j \log \log b^j)^{\frac{1}{2}}/(n \log \log n)^{\frac{1}{2}}.$$

Let $G_3(\delta, b)$ be the set of $\omega \in \Omega$ for which there is a

$$J = J_3(\delta, b, \omega) < \infty$$

such that:

$$j \geqq J \quad \text{and} \quad b^{j-1} \leqq n < b^j \quad \text{imply} \quad \|Z(n)(\cdot, \omega) - b(n, j)Z(b^j)(\cdot, \omega)\| \leqq \delta.$$

(84) Lemma. *$G_3(\delta, b) \in \mathscr{F}$. There is a $b(\delta) > 1$ such that $\mathscr{P}\{G_3(\delta, b)\} = 1$ for $1 < b < b(\delta)$.*

PROOF. Use (20). ★

To state (85), let $\eta > 0$. Let $G_4(\eta)$ be the set of $\omega \in \Omega$ for which there is an $N = N(\eta, \omega) < \infty$ such that:

$$n \geqq N \quad \text{implies} \quad d(Z_n(\cdot, \omega), K) < \eta.$$

Remember

$$d(f, K) = \inf\{d(f, g): g \in K\},$$

where g can be restricted to a countable dense subset of K without changing anything.

(85) Lemma. *$G_4(\eta) \in \mathscr{F}$ and $\mathscr{P}\{G_4(\eta)\} = 1$.*

PROOF. Choose δ with $0 < \delta < \eta$. Then determine $b(\delta) > 1$ from (84). Then choose $a > 1$ and $\varepsilon > 0$ and $\beta > 0$ and b with $1 < b < b(\delta)$, so

(86) $$a - 1 + \varepsilon + \delta + (b^{\frac{1}{2}} - 1 + \beta)(a + \varepsilon) < \eta.$$

Choose a positive integer m so large that $m\varepsilon^2 > 1$. Let

$$G = G_1(a, b, m) \cap G_2(\varepsilon, b, m) \cap G_3(\delta, b).$$

As (82–84) imply, $G \in \mathscr{F}$ and $\mathscr{P}\{G\} = 1$. I say

$$G \subset G_4(\eta).$$

To begin the argument, constrain n and j so

(87) $$b^{j-1} \leqq n < b^j.$$

Then

$$\limsup_{n,j \to \infty} b(n, j) = b^{\frac{1}{2}}.$$

Choose $J_0 < \infty$ so large that $j \geqq J_0$ implies

(88) $$b(n, j) \leqq b^{\frac{1}{2}} + \beta,$$

always provided (87). Now fix $\omega \in G$. Let $J < \infty$ be no smaller than

$$J_0, J_1(a, b, m, \omega), J_2(\varepsilon, b, m, \omega), J_3(\delta, b, \omega).$$

Let $n \geqq b^{J-1}$, and choose j to satisfy (87); so $j \geqq J$. I will show that

(89) $$d(Z_n(\cdot, \omega), K) < \eta.$$

By the (82) preliminaries,

(90) $$a^{-1} Z(b^j, m)(\cdot, \omega) \in K,$$

so (79) shows

(91) $$\|Z(b^j, m)(\cdot, \omega)\| \leqq a$$

and

(92) $\|Z(b^j, m)(\cdot, \omega) - a^{-1}Z(b^j, m)(\cdot, \omega)\| \leqq a - 1.$

By the (83) preliminaries,

(93) $$\|Z(b^j)(\cdot, \omega) - Z(b^j, m)(\cdot, \omega)\| \leqq \varepsilon.$$

By (91) and (93),

(94) $$\|Z(b^j)(\cdot, \omega)\| \leqq a + \varepsilon.$$

By (87) and the (84) preliminaries,

(95) $$\|Z(n)(\cdot, \omega) - b(n, j)Z(b^j)(\cdot, \omega)\| \leqq \delta.$$

Use (88) and (94):

(96) $$\begin{aligned}\|b(n, j)Z(b^j)(\cdot, \omega) - Z(b^j)(\cdot, \omega)\| &= (b(n, j) - 1)\|Z(b^j)(\cdot, \omega)\| \\ &\leqq (b^{\frac{1}{2}} - 1 + \beta)(a + \varepsilon).\end{aligned}$$

Add (92, 93, 95, 96):

$$\|Z(n)(\cdot, \omega) - a^{-1}Z(b^j, m)(\cdot, \omega)\| \leqq a - 1 + \varepsilon + \delta + (b^{\frac{1}{2}} - 1 + \beta)(a + \varepsilon).$$

Now (89) follows from (86) and (90). ★

Let $G_5 = \bigcap_{n=1}^{\infty} G_4(1/n)$. Then $G_5 \in \mathscr{F}$ and $\mathscr{P}\{G_5\} = 1$, by (85). But G_5 is exactly the set of $\omega \in \Omega$ with

$$\lim_{n\to\infty} d(Z_n(\cdot, \omega), K) = 0.$$

This proves part (a) of (81):

(97) Proposition. *For $\mathscr{P}$-almost all ω,*

$$\lim_{n\to\infty} d(Z_n(\cdot, \omega), K) = 0.$$

(98) Corollary. *With $\mathscr{P}$-probability* 1,

$$\limsup_{t\to\infty} |B(t)|/(2t \log\log t)^{\frac{1}{2}} \leqq 1.$$

Proof. Let $L(t) = (2t \log\log t)^{\frac{1}{2}}$. Proposition (97) and (78c) imply that $\mathscr{P}$-almost surely,

$$\limsup_{n\to\infty} \|Z_n\| \leqq 1;$$

in particular,

$$\limsup_{n\to\infty} L(n)^{-1} \max\left\{|B(ns)| : 1 - \frac{1}{n} \leqq s \leqq 1\right\} \leqq 1;$$

put $t = ns$, and see

$$\limsup_{n\to\infty} L(n)^{-1} \max\{|B(t)| : n - 1 \leqq t \leqq n\} \leqq 1.$$

As n and t approach ∞ with $n - 1 \leqq t \leqq n$,

$$L(n)/L(t) \to 1. \qquad \bigstar$$

Part (b) of the proof: motivation

Suppose $f \in K$ is rational at i/m and linear on $\left[\frac{i-1}{m}, \frac{i}{m}\right]$ for $i = 1, \ldots, m$. The set of such f, as m varies, is a (wrong) countable, dense subset of K. But for now, fix one m and one f. When is Z_n close to f? To begin with, Z_n and $Z_{n,m}$ are close, so the interesting problem is to find out when $Z_{n,m}$ is close to f. Fix a positive target error ε. Reserve $\varepsilon/2$ for the distance between Z_n and $Z_{n,m}$. What I want now is

$$\|Z_{n,m} - f\| < \varepsilon/2$$

for infinitely many n. By convexity, the norm inequality boils down to

$$\left|Z_{n,m}\left(\frac{i}{m}\right) - f\left(\frac{i}{m}\right)\right| < \varepsilon/2 \quad \text{for } i = 1, \ldots, m. \tag{99}$$

Now

$$Z_{n,m}\left(\frac{i}{m}\right) = Z_n\left(\frac{i}{m}\right) = L(n)^{-1}B\left(\frac{i}{m}n\right),$$

where

$$L(n) = (2n \log \log n)^{\frac{1}{2}}.$$

For each n, the problem is reduced to the behavior of B at a finite number of points; this is the reason for introducing $Z_{n,m}$.

I will now rewrite (99). Introduce

$$a_i = f\left(\frac{i}{m}\right) - f\left(\frac{i-1}{m}\right),$$

so

$$1 \geqq \int_0^1 f'(t)^2\, dt = m \, \Sigma_{i=1}^m a_i^2.$$

Fix positive partial errors $\delta_1, \ldots, \delta_m$, whose sum is less than $\varepsilon/2$. Then (99) follows from

$$(a_i - \delta_i)L(n) < B\left(\frac{i}{m}n\right) - B\left(\frac{i-1}{m}n\right) < (a_i + \delta_i)L(n)$$

for $i = 1, \ldots, m$. Temporarily, fix n so large that

$$2\delta_i(2m \log\log n)^{\frac{1}{2}} > (2\pi)^{\frac{1}{2}} \quad \text{for } i = 1, \ldots, m.$$

The differences $B\left(\frac{i}{m}n\right) - B\left(\frac{i-1}{m}n\right)$ are independent $N\left(0, \frac{n}{m}\right)$ as i varies. Multiply across the ith inequality by $(m/n)^{\frac{1}{2}}$ and use (4b). The ith inequality is fulfilled with probability at least

$$\exp\{-m(a_i \pm \delta_i)^2 \log\log n\},$$

where $\exp(x) = e^x$; while $(a \pm \delta)$ is $a + \delta$ when $a \geqq 0$ and $a - \delta$ when $a \leqq 0$. The system of m inequalities is fulfilled with probability at least

$$\exp[-m\,\Sigma_{i=1}^m (a_i \pm \delta_i)^2 \log\log n].$$

As n varies, the corresponding systems are dependent; the faster n increases, the weaker the dependence. The right rate is geometric: restrict n to the powers of b, for suitable $b > 1$. The system corresponding to $n = b^j$—from now on, I will call this system (j)—depends on B up to time b^j; while system $(j-1)$ depends on B up to time b^{j-1}. But system (j) depends on $\{B(t): 0 \leqq t \leqq b^{j-1}\}$ only through the first inequality, provided $b > m$. This suggests revising the first inequality to read

$$(a_1 - \delta_1)L(b^j) < B\left(\frac{1}{m}b^j\right) - B(b^{j-1}) < (a_1 + \delta_1)L(b^j),$$

and making $b > m$. The difference $B\left(\frac{1}{m}b^j\right) - B(b^{j-1})$ is normal with mean 0 and variance

$$\frac{1}{m}b^j - b^{j-1} = b^j\left(\frac{b-m}{mb}\right).$$

So the revised first inequality holds with probability at least

$$\exp\left[-\frac{mb}{b-m}(a_1 \pm \delta_1)^2 \log\log b^j\right],$$

provided j is so large that

$$2\delta_1\left[\frac{2mb}{b-m}\log\log b^j\right]^{\frac{1}{2}} > (2\pi)^{\frac{1}{2}}.$$

And the revised system holds with probability at least

$$\exp[-\theta \log\log b^j] \sim j^{-\theta},$$

where

$$\theta = m\left[\frac{b}{b-m}(a_1 \pm \delta_1)^2 + \Sigma_{i=2}^m (a_i \pm \delta_i)^2\right].$$

The sum of the probabilities is infinite, provided $\theta < 1$. To make this happen, revise C so as to include only f with

$$\int_0^1 f'(t)^2 \, dt < 1.$$

Make the δ_i so small, and then b so large, that $\theta < 1$. The revised systems (j) are independent for various j, and the sum of their probabilities is infinite. So infinitely many systems are fulfilled.

Suppose system (j) holds. Why does this make $Z(b^j, m)$ close to f? The new difficulty is a loss of control over $B\left(\frac{1}{m}b^j\right)$. But (98) shows that for large enough j,

$$\begin{aligned} |B(b^{j-1})| &\leqq 2L(b^{j-1}) \\ &\leqq 2b^{-\frac{1}{2}}L(b^j). \end{aligned}$$

This and system (j) forces $Z(b^j, m)$ to within

$$2b^{-\frac{1}{2}} + \delta_1 + \cdots + \delta_m$$

of f. For large enough b, this bigger error is also less than $\varepsilon/2$.

Come back to $\|Z_{n,m} - Z_n\|$. This is small when m is large. More exactly, if $n = b^j$ and $m\varepsilon^2/4 > 1$, this norm is eventually no more than $\varepsilon/2$ by (83). At first sight, the cause is lost: because ε and m were given. But really, only ε and f are fixed. If f has the right shape for one m, it does for all multiples of m. So you can choose m arbitrarily large.

Part (b) of the proof: formalities

For $m = 1, 2, \ldots,$ let C_m be the set of $f \in C[0, 1]$ such that:

$$f(0) = 0$$

$$f\left(\frac{i}{m}\right) \text{ is rational} \qquad \text{for } i = 1, \ldots, m$$

$$f \text{ is linear on } \left[\frac{i-1}{m}, \frac{i}{m}\right] \qquad \text{for } i = 1, \ldots, m$$

$$\int_0^1 f'(t)^2 \, dt < 1.$$

Let $C = \bigcup_{m=1}^{\infty} C_m$. Clearly, C is a countable subset of K.

(100) Lemma. *C is dense in K.*

PROOF. Fix a target $g \in K$. Let g_m agree with g at i/m for $i = 0, \ldots, m$, and be linearly interpolated. Then g_m converges to g as $m \to \infty$, because g is uniformly continuous on $[0, 1]$. And $g_m \in K$ by (80). Let $a > 1$. Then $h_a = a^{-1} g_m$ converges to g_m as $a \downarrow 1$, and $\int_0^1 h_a'(t)^2 \, dt < 1$. You should have no trouble in approximating h_a by $f \in C$. ★

To state (101), fix $\varepsilon > 0$ and $f \in C$. Let $A(f, \varepsilon)$ be the set of $\omega \in \Omega$ such that

$$\|Z_n(\cdot, \omega) - f\| < \varepsilon \quad \text{for infinitely many } n.$$

(101) Lemma. *$A(f, \varepsilon) \in \mathscr{F}$ and $\mathscr{P}\{A(f, \varepsilon)\} = 1$.*

PROOF. If $f \in C_k$, then $f \in C_{jk}$ for $j = 1, 2, \ldots$. I choose m so $f \in C_m$ and $m\varepsilon^2/4 > 1$. For $i = 1, \ldots, m$, let

$$a_i = f\left(\frac{i}{m}\right) - f\left(\frac{i-1}{m}\right),$$

so

$$1 > \int_0^1 f'(t)^2 \, dt = m \, \Sigma_{i=1}^m \, a_i^2.$$

Choose positive δ_i so small that

$$\delta_1 + \cdots + \delta_m < \varepsilon/2$$

$$m \, \Sigma_{i=1}^m \, (a_i \pm \delta_i)^2 < 1.$$

Now choose $b > m$ so large that

$$2b^{-\frac{1}{2}} + \delta_1 + \cdots + \delta_m < \varepsilon/2 \tag{102}$$

$$\theta = m\left[\frac{b}{b-m}(a_1 \pm \delta_1)^2 + \Sigma_{i=2}^m \, (a_i \pm \delta_i)^2\right] < 1. \tag{103}$$

Here,

$$\begin{aligned} a \pm \delta &= a + \delta \quad \text{when } a \geqq 0 \\ &= a - \delta \quad \text{when } a < 0. \end{aligned}$$

Let

$$L(n) = (2n \log \log n)^{\frac{1}{2}}.$$

Introduce the events $F_{i,j}$ for $i = 1, \ldots, m$:

$$F_{1,j} = \left\{(a_1 - \delta_1)L(b^j) < B\left(\frac{1}{m} b^j\right) - B(b^{j-1}) < (a_1 + \delta_1)L(b^j)\right\};$$

and for $i = 2, \ldots, m$,

$$F_{i,j} = \left\{(a_i - \delta_i)L(b^j) < B\left(\frac{i}{m}b^j\right) - B\left(\frac{i-1}{m}b^j\right) < (a_i + \delta_i)L(b^j)\right\}.$$

Let $E_j = \bigcap_{i=1}^m F_{i,j}$.

For fixed j, the sets $F_{1,j}, \ldots, F_{m,j}$ are independent by (11c), so

$$\mathscr{P}\{E_j\} = \Pi_{i=1}^m \mathscr{P}\{F_{i,j}\}.$$

By (4b), an $N(0, 1)$ random variable falls in the interval (x, y) with probability at least $\exp\{-\frac{1}{2}\max(x^2, y^2)\}$, provided $y - x > (2\pi)^{\frac{1}{2}}$. Suppose j is so large that

$$2\delta_1\left(\frac{2mb}{b-m}\log\log b^j\right)^{\frac{1}{2}} > (2\pi)^{\frac{1}{2}}$$

$$2\delta_i(2m\log\log b^j)^{\frac{1}{2}} > (2\pi)^{\frac{1}{2}} \quad \text{for } i = 2, \ldots, m.$$

Then

$$\mathscr{P}\{F_{1,j}\} \geqq \exp\left[-\frac{mb}{b-m}(a_1 \pm \delta_1)^2 \log\log b^j\right]$$

$$\mathscr{P}\{F_{i,j}\} \geqq \exp\left[-m(a_i \pm \delta_i)^2 \log\log b^j\right] \quad \text{for } i = 2, \ldots, m;$$

because

$$B\left(\frac{1}{m}b^j\right) - B(b^{j-1}) \quad \text{is} \quad N\left(0, \frac{b-m}{mb}b^j\right)$$

$$B\left(\frac{i}{m}b^j\right) - B\left(\frac{i-1}{m}b^j\right) \quad \text{is} \quad N\left(0, \frac{1}{m}b^j\right).$$

Consequently,

$$\mathscr{P}\{E_j\} \geqq \exp\{-\theta\log\log b^j\} \sim j^{-\theta},$$

where

$$\theta = m\left[\frac{b}{b-m}(a_1 \pm \delta_1)^2 + \Sigma_{i=2}^m (a_i \pm \delta_i)^2\right]$$

is less than 1 by (103). Therefore,

$$\Sigma_{j=1}^\infty \mathscr{P}\{E_j\} = \infty.$$

Use (11c) to check that $E_1, E_2, \ldots$ are independent; the definition of $F_{1,j}$ was chosen to make this work. Now Borel-Cantelli guarantees

$$\mathscr{P}\{\limsup E_j\} = 1.$$

Introduce new events $F^*_{i,j}$ for $i = 1, \ldots, m$:

$$F^*_{1,j} = \left\{(a_1 - \delta_1 - 2b^{-\frac{1}{2}})L(b^j) < B\left(\frac{1}{m}b^j\right) < (a_1 + \delta_1 + 2b^{-\frac{1}{2}})L(b^j)\right\};$$

and for $i = 2, \ldots, m$,

$$F^*_{i,j} = F_{i,j}.$$

Let $E^*_j = \bigcap_{i=1}^m F^*_{i,j}$. By (98), for $\mathscr{P}$-almost all ω there is a $J = J(b, \omega) < \infty$ such that $j > J$ implies

$$|B(b^{j-1}, \omega)| \leqq 2b^{-\frac{1}{2}}L(b^j);$$

then $\omega \in E_j$ implies $\omega \in E^*_j$. Therefore,

$$\mathscr{P}\{\limsup E^*_j\} = 1.$$

I say

(104) $$E^*_j \subset \{\|Z(b^j, m) - f\| < \varepsilon/2\}.$$

To begin with,

$$\begin{aligned} \|Z(b^j, m) - f\| &= \max_i \left| Z(b^j, m)\left(\frac{i}{m}\right) - f\left(\frac{i}{m}\right)\right| \\ &= \max_i \left| Z(b^j)\left(\frac{i}{m}\right) - f\left(\frac{i}{m}\right)\right| \\ &\leqq \sum_{i=1}^m \left| Z(b^j)\left(\frac{i}{m}\right) - Z(b^j)\left(\frac{i-1}{m}\right) - a_i\right|; \end{aligned}$$

use the convexity in line 1, the definition of $Z(n, m)$ in line 2, and the definition of a_i in line 3. On E^*_j:

$$\left| Z(b^j)\left(\frac{i}{m}\right) - Z(b^j)\left(\frac{i-1}{m}\right) - a_i\right| = L(b^j)^{-1}\left| B\left(\frac{i}{m}b^j\right) - B\left(\frac{i-1}{m}b^j\right) - a_iL(b^j)\right|$$

is at most $\delta_1 + 2b^{-\frac{1}{2}}$ for $i = 1$, and δ_i for $i = 2, \ldots, m$. Now (104) follows from (102).

Consequently, $\mathscr{P}$-almost surely,

$$\|Z(b^j, m) - f\| < \varepsilon/2$$

for infinitely many j. Use (83) and the condition $m\varepsilon^2/4 > 1$ to see that $\mathscr{P}$-almost surely,

$$\|Z(b^j) - Z(b^j, m)\| \leqq \varepsilon/2$$

for all sufficiently large j. The conclusion is that $\mathscr{P}$-almost surely,

$$\|Z(b^j) - f\| < \varepsilon \quad \text{for infinitely many } j.$$ ★

Let

$$A = \bigcap \{A(f, 1/n) : f \in C \text{ and } n = 1, 2, \ldots\}.$$

As (101) implies, $A \in \mathscr{F}$ and $\mathscr{P}(A) = 1$. But A is the set of $\omega \in \Omega$ such that

$$\liminf_{n\to\infty} \|Z_n(\cdot, \omega) - f\| = 0$$

for all $f \in C$. This proves part (b) of (81):

(105) Proposition. *For $\mathscr{P}$-almost all ω:*

$$\liminf_{n\to\infty} \|Z_n(\cdot, \omega) - f\| = 0$$

for all $f \in C$, which is a countable dense subset of K.

The climax

PROOF OF (81). Use (97), (105), and (76). ★

The classical iterated logarithm

Khintchine's (1924) law of the iterated logarithm follows, although this derivation is rather silly; for an elegant one, see (Lévy, 1965, pp. 226–228).

(106) Theorem. *With $\mathscr{P}$-probability* 1,

$$\limsup_{t\to\infty} \frac{B(t)}{\sqrt{2t \log\log t}} = 1$$

$$\liminf_{t\to\infty} \frac{B(t)}{\sqrt{2t \log\log t}} = -1.$$

PROOF. Let $f(x) = x$ for $0 \leqq x \leqq 1$. Then $f(1) = 1$ and $f \in K$. So for $\mathscr{P}$-almost all ω, there is a subsequence $n'(\omega)$ such that

$$Z_{n'(\omega)}(\cdot, \omega) \to f.$$

For such an ω,

$$B(n'(\omega), \omega)/(2n'(\omega) \log\log n'(\omega))^{\frac{1}{2}} = Z_{n'(\omega)}(1, \omega) \to 1,$$

so

$$\limsup_{t\to\infty} \frac{B(t, \omega)}{\sqrt{2t \log\log t}} \geqq 1.$$

The opposite inequality comes from (98). The lim inf is symmetric. ★

(107) Corollary. *With $\mathscr{P}$-probability* 1,

$$\limsup_{t \downarrow 0} \frac{B(t)}{\sqrt{2t \log \log (1/t)}} = 1$$

$$\liminf_{t \downarrow 0} \frac{B(t)}{\sqrt{2t \log \log (1/t)}} = -1.$$

PROOF. Use (106), and the duality between 0 and ∞ expressed in (13). ★

6. THE SKOROKHOD REPRESENTATION

Throughout this section, let F be a distribution function on the real line, with mean 0: that is,

$$\int_{-\infty}^{\infty} x \, F(dx) = 0.$$

As (108) will show, F can be represented as an average of two-point, mean 0 distributions. More precisely, let $G(0, 0)$ be the distribution function assigning mass 1 to 0. For $u > 0$ and $v > 0$, let $G(u, v)$ be the unique distribution function which assigns measure 1 to $\{-u, v\}$ and has mean 0: namely, $G(u, v)$ assigns mass $v/(u + v)$ to $-u$ and $u/(u + v)$ to v. Clearly, $G(u, v)$ has mean square uv: that is,

$$\int_{-\infty}^{\infty} x^2 \, G(u, v)(dx) = uv.$$

(108) Lemma. *There are nonnegative random variables U and V on a probability triple $(\mathscr{X}, \Sigma, m)$ such that: $U = 0$ iff $V = 0$ and*

$$F(x) = \int_{t \in \mathscr{X}} G[U(t), V(t)](x) \, m(dt).$$

You can choose $\mathscr{X}$ as an interval and Σ as the Borel subsets of $\mathscr{X}$.

PROOF. If F assigns mass $\alpha > 0$ to 0, then $F = \alpha G(0, 0) + (1 - \alpha)F^*$, where the distribution function F^* assigns mass 0 to 0 and has mean 0. It is therefore permissible to assume that F assigns mass 0 to 0. Let $f = F^{-1}$, a nondecreasing function on the Lebesgue unit interval with distribution function F. Let

$$z = \inf \{x : f(x) > 0\} = \sup \{x : f(x) < 0\},$$

the two being equal because $\{x : f(x) = 0\}$ is empty or a singleton. Clearly, $0 < z < 1$. Normalize f so $f(z) = 0$, and f is continuous from the right on

$(z, 1)$, and f is continuous from the left on $(0, z)$. The best case to think about is continuous f. For $z \leqq x \leqq 1$, let

$$A(x) = \int_z^x f(t)\,dt.$$

For $0 \leqq y \leqq z$, let

$$C(y) = \int_y^z f(t)\,dt.$$

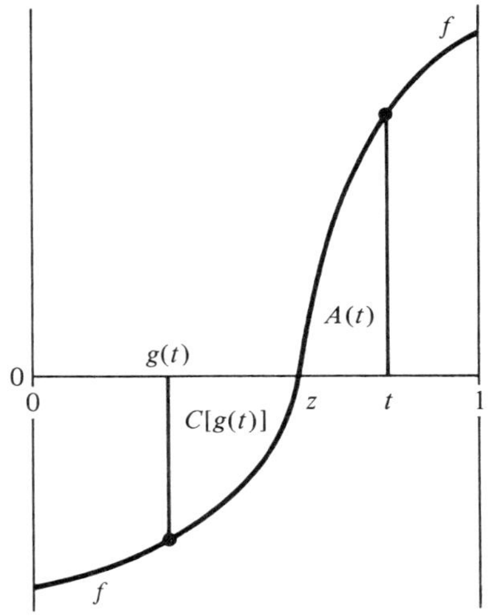

Figure 8.

The functions A and C are illustrated in Figure 8. Clearly, $A(z) = C(z) = 0$. Moreover, A and C are continuous and strictly increasing. And

$$C(0) + A(1) = \int_0^1 f(t)\,dt = \int_{-\infty}^{\infty} x\,F(dx) = 0.$$

Consequently, for each $t \in [z, 1]$ there is a unique $g(t) \in [0, z]$ such that

$$A(t) + C[g(t)] = 0. \tag{109}$$

The function g is illustrated in Figure 8. You should check that g is continuous and strictly decreasing, $g(z) = z$, and $g(1) = 0$. Let $^+$ denote differentiation from the right, and let $h(t) = f(t)/|f(g(t))|$. If f is continuous, $^+$ can be replaced by the usual derivative. Differentiate (109) from the right:

$$g^+(t) = -h(t) \quad \text{on } (z, 1). \tag{110}$$

Use (3.69) and continuity:

$$\text{(111)} \qquad \int_z^1 h(t)\,dt = g(z) - g(1) = z.$$

And g is absolutely continuous, with density $-h$.

Let $\mathscr{X}$ be $(z, 1)$; let Σ be the Borel subsets of $\mathscr{X}$; and let m be the measure on Σ whose density with respect to Lebesgue measure is $1 + h$ on $\mathscr{X}$. As (111) implies, $m\{\mathscr{X}\} = 1$. Let $U(t) = |f(g(t))|$ and $V(t) = f(t)$ for $t \in \mathscr{X}$. In particular $U(t) > 0$ and $V(t) > 0$. Let $G(t) = G[U(t), V(t)]$. Let $\pi(t)$ be the mass $G(t)$ assigns to $-U(t)$, namely $h(t)/[1 + h(t)]$. I have to argue

$$\int G(t)\, m(dt) = F.$$

Let ϕ be a bounded, measurable function on $(-\infty, \infty)$, which vanishes on $(0, \infty)$. You should check the following computation, by making the change of variables $s = g(t)$ and using the fact that g has density $-h$.

$$\begin{aligned}
\int_{t\in\mathscr{X}} \int_{-\infty<y<\infty} \phi(y)\, G(t)(dy)\, m(dt) &= \int_{t\in\mathscr{X}} \phi[-U(t)]\, \pi(t)\, m(dt) \\
&= \int_{t\in(z,1)} \phi[-U(t)]\, \pi(t) \frac{dm}{dt}\, dt \\
&= \int_{t\in(z,1)} \phi\{f[g(t)]\} \frac{h(t)}{1 + h(t)} [1 + h(t)]\, dt \\
&= \int_{s\in(0,z)} \phi\{f(s)\}\, ds \\
&= \int_{s\in(0,1)} \phi\{f(s)\}\, ds \\
&= \int_{-\infty}^{\infty} \phi(x)\, F(dx).
\end{aligned}$$

The argument for ϕ which vanish on $(-\infty, 0]$ is easier. ★

To state (112), let $(\Omega, \mathscr{F}, \mathscr{P})$ be a probability triple. For each $\omega \in \Omega$, let $G(\omega)(\cdot)$ be a distribution function on $(-\infty, \infty)$. Suppose $\omega \to G(\omega)(x)$ is $\mathscr{F}$-measurable for each $x \in (-\infty, \infty)$. Let

$$H(x) = \int_\Omega G(\omega)(x)\, \mathscr{P}(d\omega).$$

Let f be a nonnegative, measurable function on $(-\infty, \infty)$. Then

(112) Lemma. **(a)** *H is a distribution function.*

(b) $$\int_\Omega \int_{-\infty}^{\infty} f(x)\, G(\omega)(dx)\, \mathscr{P}(d\omega) = \int_{-\infty}^{\infty} f(x)\, H(dx).$$

PROOF. *Claim* (*a*) is easy, using dominated convergence.

Claim (*b*) is easy when f is the indicator of a half-line $(-\infty, a]$. Now extend, using linearity and monotone convergence. ★

Remember that $C[0, \infty)$ is the space of continuous, real-valued functions f on $[0, \infty)$. For $t \in [0, \infty)$, the coordinate $\xi(t)$ is defined by

$$\xi(t, f) = f(t).$$

And $C[0, \infty)$ is given the smallest σ-field Σ over which all the $\xi(t)$ are measurable. Let S be the sets of pairs (u, v) of nonnegative real numbers, such that $u = 0$ iff $v = 0$. Give S the Borel σ-field $\mathscr{B}$. Define a function T on $S \times C[0, \infty)$ as follows: $T(u, v, f)$ is the least t if any with $f(t) = -u$ or v; and $T(u, v, f)$ is ∞ if there is no such t.

(113) Lemma. *T is measurable, that is, relative to $\mathscr{B} \times \Sigma$.*

PROOF. $T(u, v, f) > t$ iff there are rational numbers a and b with

$$-u < a < 0 < b < v$$

and $a \leqq f(r) \leqq b$ for all rational r in $[0, t]$. ★

I now want to return to normalized Brownian motion B on $(\Omega, \mathscr{F}, \mathscr{P})$. Write E for expectation with respect to $\mathscr{P}$. Let u and v be positive numbers. Let τ be the least t with $B(t)$ either $-u$ or v, relying on (18). Thus

$$\tau = T(u, v, B).$$

(114) Lemma. **(a)** $$\mathscr{P}\{B(\tau) = -u\} = \frac{v}{u+v}$$

(b) $$\mathscr{P}\{B(\tau) = v\} = \frac{u}{u+v}$$

(c) $$E\{\tau\} = uv.$$

PROOF. The process $\{B(t)\}$ is a martingale, so (3.29) makes

$$E\{B(\tau)\} = 0.$$

Since $B(\tau)$ is either $-u$ or v, claims (a) and (b) are vindicated. In particular,

$$E\{B(\tau)^2\} = uv.$$

The process $\{B(t)^2 - t\}$ is a martingale, so (3.29) makes

$$E\{B(\tau)^2 - \tau\} = 0.$$

That is,

$$E\{\tau\} = E\{B(\tau)^2\} = uv,$$

proving (c). You can use (16) to check (3.29b). ★

Remember that F is a distribution function on $(-\infty, \infty)$, with mean 0. And B is normalized Brownian motion on $(\Omega, \mathscr{F}, \mathscr{P})$. Suppose there are nonnegative random variables U and V on $(\Omega, \mathscr{F}, \mathscr{P})$, such that (U, V) is independent of B, and $U = 0$ iff $V = 0$, and $E\{G(U, V)\} = F$.

(115) Lemma. **(a)** $T(U, V, B) < \infty$.

(b) $E\{T(U, V, B)\} = \int x^2 \, F(dx)$.

(c) *The distribution function of* $B[T(U, V, B)]$ *is* F.

Pedantry. $B[T(U, V, B)](\omega) = B\{T[U(\omega), V(\omega), B(\cdot\,, \omega)], \omega\}$.

PROOF. *Claim* (*a*) follows from (18).

Claims (*b*) *and* (*c*). Fix a real number x. Let

$$g(u, v) = E\{T(u, v, B)\}$$

$$h(u, v) = \mathscr{P}\{B[T(u, v, B)] \leqq x\}.$$

By (114):

$$\begin{aligned} g(u, v) &= uv \\ &= \int_{-\infty}^{\infty} y^2 \, G(u, v)(dy); \\ h(u, v) &= G(u, v)(x). \end{aligned}$$

Use Fubini, with (113) to do the measurability and (112) to interchange integrals:

$$\begin{aligned} E\{T(U, V, B)\} &= E\{g(U, V)\} \\ &= E\left\{\int_{-\infty}^{\infty} y^2 \, G(U, V)(dy)\right\} \\ &= \int_{-\infty}^{\infty} y^2 \, E\{G(U, V)\}(dy) \\ &= \int_{-\infty}^{\infty} y^2 \, F(dy); \end{aligned}$$

$$\begin{aligned}\mathscr{P}\{B[T(U,V,B)] \leqq x\} &= E\{h(U,V)\}\\ &= E\{G(U,V)(x)\}\\ &= F(x).\end{aligned}$$ ★

For (117), keep F a distribution function on $(-\infty, \infty)$, with mean 0. Do not assume that F has finite mean square. And B is still normalized Brownian motion on $(\Omega, \mathscr{F}, \mathscr{P})$. Suppose $(U_1, V_1), (U_2, V_2), \ldots$ are independent and identically distributed random vectors on $(\Omega, \mathscr{F}, \mathscr{P})$, independent of B. For all n, suppose $U_n \geqq 0$ and $V_n \geqq 0$ and $U_n = 0$ iff $V_n = 0$, and $E\{G(U_n, V_n)\} = F$.

(116) Remark. You can create such a system by appealing to (6), (108), and the existence of product measure.

The next result is due to Skorokhod (1965); I will call it the *Skorokhod representation* for F.

(117) Theorem. *There are nonnegative random variables* $\tau_1 \leqq \tau_2 \leqq \tau_3 \leqq \cdots$ *on* $(\Omega, \mathscr{F}, \mathscr{P})$ *such that:*

(a) $\tau_1, \tau_2 - \tau_1, \tau_3 - \tau_2, \ldots$ *are independent, identically distributed and finite;*

(b) $E\{\tau_1\} = \int_{-\infty}^{\infty} x^2 \, F(dx)$;

(c) $B(\tau_1), B(\tau_2) - B(\tau_1), B(\tau_3) - B(\tau_2), \ldots$ *are independent and identically distributed;*

(d) $B(\tau_1)$ *has distribution function* F.

NOTE. By (a), (b), and the strong law,

(e) $\tau_n/n \to \int_{-\infty}^{\infty} x^2 \, F(dx)$ *almost surely.*

PROOF. Let $\tau_0 = 0$. Suppose τ_n constructed. Let τ_{n+1} be the least $t \geqq \tau_n$ with

$$B(t) = B(\tau_n) - U_{n+1} \quad \text{or} \quad B(t) = B(\tau_n) + V_{n+1}.$$

Properties (b) and (d) are immediate from (115), because

$$\tau_1 = T(U_1, V_1, B) \quad \text{and} \quad B(\tau_1) = B[T(U_1, V_1, B)].$$

I will now use strong Markov to argue that $\tau_{n+1} - \tau_n$ is distributed like τ_1 and is independent of $\tau_1, \ldots, \tau_n$; while $B(\tau_{n+1}) - B(\tau_n)$ is distributed like $B(\tau_1)$ and is independent of $B(\tau_1), \ldots, B(\tau_n)$. Let $\mathscr{A}_n(t)$ be the σ-field spanned by $B(s)$ for $0 \leqq s \leqq t$ and by $(U_1, V_1), \ldots, (U_n, V_n)$. Clearly, $\{\mathscr{A}_n(t): 0 \leqq t < \infty\}$ is admissible; and $\mathscr{A}_n(t)$ increases with n.

Call σ a *strict* Markov time relative to $\mathscr{A}_n(\cdot)$ iff $\{\tau_n \leqq t\} \in \mathscr{A}_n(t)$ for all t. Then $\mathscr{A}_n(\sigma)$ is by definition the σ-field of all measurable sets A with $A \cap \{\sigma \leqq t\} \in \mathscr{A}_n(t)$ for all t. Check that σ is a Markov time, and $\mathscr{A}_n(\sigma) \subset \mathscr{A}_n(\sigma+)$. I will argue by induction that

(α) τ_n is a strict Markov time relative to $\mathscr{A}_n(\cdot)$

(β) $\tau_1, \ldots, \tau_n$ and $B(\tau_1), \ldots, B(\tau_n)$ are $\mathscr{A}_n(\tau_n)$-measurable.

These assertions are trivial for $n = 0$. Suppose them for $n = m$. Now

$$\tau_{m+1} \leqq t \quad \text{and} \quad B(\tau_{m+1}) - B(\tau_m) > y > 0$$

iff: $V_{m+1} > y$; and for every rational $r < V_{m+1}$, there is a rational s with

$$\tau_m \leqq s \leqq t \quad \text{and} \quad B(s) > B(\tau_m) + r.$$

By inductive assumption: $\{\tau_m \leqq s\} \in \mathscr{A}_m(s) \subset \mathscr{A}_{m+1}(t)$; and $B(\tau_m)$ retracted to $\{\tau_m \leqq s\}$ is measurable relative to $\mathscr{A}_m(s)$, therefore relative to $\mathscr{A}_{m+1}(t)$. So

$$\{\tau_{m+1} \leqq t \text{ and } B(\tau_{m+1}) - B(\tau_m) > y > 0\} \in \mathscr{A}_{m+1}(t).$$

Similarly,

$$\{\tau_{m+1} \leqq t \text{ and } B(\tau_{m+1}) - B(\tau_m) < y < 0\} \in \mathscr{A}_{m+1}(t).$$

It's easy to get

$$\{\tau_{m+1} \leqq t \text{ and } B(\tau_{m+1}) - B(\tau_m) = 0\} \in \mathscr{A}_{m+1}(t).$$

This proves (α) for $n = m + 1$, and also shows $B(\tau_{m+1})$ is $\mathscr{A}_{m+1}(\tau_{m+1})$-measurable. Clearly, τ_{m+1} is $\mathscr{A}_{m+1}(\tau_{m+1})$-measurable. You can check

$$\mathscr{A}_n(\tau_n) \subset \mathscr{A}_{m+1}(\tau_{m+1}) \quad \text{for } n \leqq m,$$

because $\mathscr{A}_n(t)$ and τ_n nondecrease with n. This finishes (β).

Let

$$Y_{n+1}(t) = B(\tau_n + t) - B(\tau_n).$$

Clearly, Y_{n+1} and $\mathscr{A}_n(\tau_n)$ are measurable on B and $(U_1, V_1), \ldots, (U_n, V_n)$. Therefore, (U_{n+1}, V_{n+1}) is independent of the σ-field spanned by Y_{n+1} and $\mathscr{A}_n(\tau_n)$. As strong Markov (21) implies, Y_{n+1} is normalized Brownian motion and is independent of $\mathscr{A}_n(\tau_n)$. Consequently, the three objects

$$(U_{n+1}, V_{n+1}), \quad Y_{n+1}, \quad \text{and} \quad \mathscr{A}_n(\tau_n)$$

are mutually independent. Therefore, the triple $(U_{n+1}, V_{n+1}, Y_{n+1})$ is distributed like the triple (U_1, V_1, B), and is independent of $\mathscr{A}_n(\tau_n)$. Next

$$\tau_{n+1} - \tau_n = T(U_{n+1}, V_{n+1}, Y_{n+1})$$

and

$$B(\tau_{n+1}) - B(\tau_n) = Y_{n+1}[T(U_{n+1}, V_{n+1}, Y_{n+1})].$$

Consequently, the pair

$$(\tau_{n+1} - \tau_n, B(\tau_{n+1}) - B(\tau_n))$$

is distributed like the pair $(\tau_1, B(\tau_1))$ and is independent of $\mathscr{A}_n(\tau_n)$. ★

The next two results (118–119) were used in Chapter 3 of *MC*; they will not be used elsewhere in the book. To state them, let F be a distribution function on $(-\infty, \infty)$, with mean 0 and positive finite variance σ^2:

$$\int_{-\infty}^{\infty} x\, F(dx) = 0$$

$$0 < \int_{-\infty}^{\infty} x^2\, F(dx) = \sigma^2 < \infty.$$

On some probability triple $(\mathscr{X}, \Sigma, \mu)$, let $X_1, X_2, \ldots$ be independent, with common distribution function F. Let $S_0 = 0$, and for $n \geqq 1$ let

$$S_n = X_1 + \cdots + X_n.$$

(118) Corollary. *Let* $\varepsilon > 0$ *and* $r > 1$. *Let* $\mu(r, n) = \mu(A)$, *where* A *is the event that*

$$\varepsilon n^{\frac{1}{2}} < \max \{|S_j - S_k| : 0 \leqq j \leqq n \text{ and } j \leqq k \leqq rj\}.$$

Then

$$\lim_{r \downarrow 1} \limsup_{n \to \infty} \mu(r, n) = 0.$$

PROOF. Let $\tau_1, \tau_2, \ldots$ be the times in the Skorokhod representation (117) for F. Let $\hat{S}_n = B(\tau_n)$. The $\mathscr{P}$-distribution of $(\hat{S}_1, \hat{S}_2, \ldots)$ coincides with the μ-distribution of $(S_1, S_2, \ldots)$, so I only have to prove (118) with $\hat{S}$ in place of S. Let M be a positive integer. Introduce the following events.

$$C = \{\varepsilon n^{\frac{1}{2}} < \max [|\hat{S}_j - \hat{S}_k| : M \leqq j \leqq n \text{ and } j \leqq k \leqq rj]\}$$

$$D = \left\{ r - 1 > \left| \frac{\tau_m}{m} - \sigma^2 \right| \text{ for all } m \geqq M \right\}$$

$$E = \left\{ \varepsilon n^{\frac{1}{2}} < \max \left[|B(t) - B(s)| : 0 \leqq s \leqq n(\sigma^2 + r - 1) \text{ and } s \leqq t \leqq r \frac{\sigma^2 + r - 1}{\sigma^2 - r + 1} s \right] \right\}$$

$$H = \{\varepsilon n^{\frac{1}{2}} < \max [|\hat{S}_j - \hat{S}_k| : 0 \leqq j \leqq M \text{ and } j \leqq k \leqq rj]\}.$$

By analysis,

$$A \subset C \cup H \quad \text{and} \quad C \cap D \subset E.$$

By logic,

$$C = (C \cap D) \cup (C \setminus D).$$

So

$$C \subset E \cup (\Omega \setminus D).$$

Therefore,

$$\mu(r, n) = \mathscr{P}\{A\} \leqq \mathscr{P}\{E\} + \mathscr{P}\{\Omega \setminus D\} + \mathscr{P}\{H\}.$$

Fix a positive target error δ. By (19), there is an $r(\delta) > 1$ so small that $\mathscr{P}\{E\} < \delta/3$ for $1 < r < r(\delta)$. Fix any such r. Using (117e), choose $M = M(r, \delta) < \infty$ so large that $\mathscr{P}\{\Omega \setminus D\} < \delta/3$. Now choose

$$N = N(M, r, \delta) < \infty$$

so large that $n \geqq N$ implies $\mathscr{P}\{H\} < \delta/3$. Then

$$\mu(r, n) < \delta \quad \text{for } n \geqq N.$$

★

(119) Corollary. *Let* $\varepsilon > 0$. *There is an* $r > 1$, *which depends only on* ε/σ, *such that* $\mu(\limsup A_n) = 0$, *where* A_n *is the event that*

$$\max\{|S_j - S_k| : 0 \leqq j \leqq n \text{ and } j \leqq k \leqq rj\}$$

exceeds $\varepsilon(n \log \log n)^{\frac{1}{2}}$.

PROOF. Use (20) and the Skorokhod representation (117), as in the proof of (118).

★

7. DONSKER'S INVARIANCE PRINCIPLE

Throughout this section, F is a distribution function on $(-\infty, \infty)$, with mean 0 and positive finite variance σ^2:

$$\int_{-\infty}^{\infty} x\, F(dx) = 0 \quad \text{and} \quad 0 < \int_{-\infty}^{\infty} x^2\, F(dx) = \sigma^2 < \infty.$$

Let $X_1, X_2, \ldots$ be independent random variables on some convenient probability triple $(\mathscr{X}, \mathscr{A}, \mu)$, with common distribution function F. Let $S_0 = 0$ and

$$S_n = X_1 + \cdots + X_n.$$

Remember that B is normalized Brownian motion on the triple $(\Omega, \mathscr{F}, \mathscr{P})$; convention (18) applies, and $(\Omega, \mathscr{F}, \mathscr{P})$ supports enough variables independent of B for (117) to hold. The construct exists by (116). You should review the *preliminaries* of Section 5, and remember that $C[0, 1]$ is the complete, separable metric space of continuous functions on $[0, 1]$, with the sup metric d and the Borel σ-field Σ generated by the d-open sets.

For $\sigma > 0$, let M_σ map Ω into $C[0, 1]$ as follows:

$$(M_\sigma\omega)(t) = B(\sigma^2 t, \omega) \quad \text{for } \omega \in \Omega \text{ and } t \in [0, 1].$$

Of course, M_σ is measurable. Let $\pi_\sigma = \mathscr{P}M_\sigma^{-1}$, a probability on $(C[0, 1], \Sigma)$. Thus, π_σ is the distribution of Brownian motion with variance parameter σ^2, starting from 0, and time parameter in $[0, 1]$.

Suppose $\{s_n : n = 0, 1, \ldots\}$ is a sequence of real numbers. Define $s_{(n)} \in C[0, 1]$ by the two requirements:

$$s_{(n)}(i/n) = n^{-\frac{1}{2}} s_i \qquad \text{for } i = 0, \ldots, n;$$

$$s_{(n)} \quad \text{is linear on} \quad \left[\frac{i-1}{n}, \frac{i}{n}\right] \qquad \text{for } i = 1, \ldots, n.$$

If $x \in X$, then $\{S_n(x) : n = 0, 1, \ldots\}$ is a sequence of real numbers, so

$$S_{(n)}(\cdot, x) = S_{(n)}(x) \in C[0, 1]$$

is defined. This process was illustrated in Figure 3. You should check that $S_{(n)}$ is a measurable mapping from $(\mathscr{X}, \mathscr{A})$ to $(C[0, 1], \Sigma)$, using (73a).

Here is Donsker's (1951) invariance principle; the proof is deferred, for the sake of examples and lemmas.

(120) Theorem. *Let φ be a bounded, real-valued, measurable function on $C[0, 1]$, which is continuous π_σ-almost everywhere. Then*

$$\lim_{n\to\infty} \int_{\mathscr{X}} \varphi(S_{(n)})\, d\mu = \int_{C[0,1]} \varphi\, d\pi_\sigma.$$

Here are some examples to illustrate (120). Recall that $\xi(t)$ is the tth coordinate function on $C[0, 1]$, for $0 \leqq t \leqq 1$:

$$\xi(t, f) = f(t).$$

(121) Example. The μ-distribution of $n^{-\frac{1}{2}} S_n$ converges to the π_σ-distribution of $\xi(1)$.

NOTE. The π_σ-distribution of $\xi(1)$ is $N(0, \sigma^2)$; so (121) is the classical central limit theorem.

PROOF. Of course, $\xi(1)$ is continuous and $\xi(1) \circ S_{(n)} = n^{-\frac{1}{2}} S_n$. Let ϕ be a bounded, continuous function on the line. Then $\phi \circ \xi(1)$ is a bounded, continuous function on $C[0, 1]$. And

$$\int_{\mathscr{X}} \phi(n^{-\frac{1}{2}} S_n)\, d\mu = \int_{\mathscr{X}} \phi \circ \xi(1) \circ S_{(n)}\, d\mu$$

$$\to \int_{C[0,1]} \phi \circ \xi(1)\, d\pi_\sigma. \qquad \bigstar$$

(122) Example. Let $\alpha > 0$. The μ-distribution of

$$Z_n = n^{-\alpha/2} \max \{|S_j|^\alpha : 1 \leqq j \leqq n\}$$

converges to the π_σ-distribution of

$$\zeta = \max \{|\xi(t)|^\alpha : 0 \leqq t \leqq 1\}.$$

NOTE. The π_σ-distribution of ζ is given by (34).

PROOF. Of course, ζ is continuous and $Z_n = \zeta \circ S_{(n)}$. Let ϕ be a bounded, continuous function on the line. Then $\phi \circ \zeta$ is a bounded, continuous function on $C[0, 1]$. And

$$\int_{\mathscr{X}} \phi(Z_n)\, d\mu = \int_{\mathscr{X}} \phi \circ \zeta \circ S_{(n)}\, d\mu$$

$$\to \int_{C[0,1]} \phi \circ \zeta\, d\pi_\sigma. \qquad \bigstar$$

(123) Example. Let s_j be 1 or 0 according as S_j is positive or nonpositive. Let

$$L_n = \frac{1}{n} \Sigma_{j=1}^n s_j.$$

Let

$$\lambda = \text{Lebesgue} \{t : 0 \leqq t \leqq 1 \text{ and } \xi(t) > 0\}.$$

The μ-distribution of L_n converges to the π_σ-distribution of λ.

NOTE. The π_σ-distribution of λ is arcsine (Itô and McKean, 1965, p. 57): its density is

$$x \to \frac{1}{\pi\sqrt{x(1-x)}} \quad \text{for} \quad 0 < x < 1.$$

PROOF. The function λ is measurable. Although λ is not continuous, it is continuous at $f \in C[0, 1]$ such that

$$\text{Lebesgue} \{t : 0 \leqq t \leqq 1 \text{ and } f(t) = 0\} = 0.$$

This set has π_σ-probability 1, by (57c). Let ϕ be a bounded, continuous function on the line. Then $\phi \circ \lambda$ is bounded, measurable, and continuous π_σ-almost everywhere. Let $e_j = 0$ unless: $S_j > 0$ and $S_{j-1} < 0$, or $S_j \leqq 0$ and $S_{j-1} > 0$, in which cases $e_j = 1$. Now

$$\int_{\mathscr{X}} \phi \circ \lambda \circ S_{(n)}\, d\mu \to \int_{C[0,1]} \phi \circ \lambda\, d\pi_\sigma,$$

and

$$|\lambda \circ S_{(n)} - L_n| \leqq \frac{1}{n} \Sigma_{j=1}^n e_j \to 0$$

with μ-probability 1.

To pin the last point down, let E be the set of $n = 1, 2, \ldots,$ such that: either $|S_n| \leqq M$ or $|X_n| > M$. Let

$$E_n = E \cap \{1, \ldots, n\},$$

and let $\# E_n$ be the number of elements of E_n. By $(MC, 3.17)$ and the strong law,

$$\lim_{n\to\infty} \# E_n/n = \mu\{|X_1| > M\}$$

with μ-probability 1. Clearly, $e_n = 0$ for $n \notin E$, so

$$\limsup_{n\to\infty} \frac{1}{n} \Sigma_{m=1}^n e_m \leqq \mu\{|X_1| > M\}$$

with μ-probability 1. Let $M \uparrow \infty$. ★

(124) Example. Let C_n be the last change of sign before n. More explicitly: if $S_n > 0$, let C_n be the largest $j < n$ with $S_j \leqq 0$; if $S_n \leqq 0$, let C_n be the largest $j < n$ with $S_j > 0$, and $C_n = \infty$ if no such j exists. Let $L_n = C_n/n$. If $\xi(1) > 0$, let λ be the largest $s < 1$ with $\xi(s) \leqq 0$, and $\lambda = \infty$ if no such s exists. If $\xi(1) \leqq 0$, let λ be the sup of $s < 1$ with $\xi(s) > 0$, and $\lambda = \infty$ if no such s exists. Usually, $\lambda(f)$ is the last zero of f in $[0, 1]$. Then the μ-distribution of L_n converges to the π_σ-distribution of λ.

NOTE. The π_σ-distribution of this λ is also arcsine. You can prove this fairly easily, using (29) and (11).

PROOF. The function λ is measurable. Although λ is not continuous, it is continuous at the $f \in C[0, 1]$ which satisfy:

$f(0) = 0$ and $f(1) \neq 0$ and $\lambda(f) < \infty$;

f is positive at points arbitrarily close to $\lambda(f)$ on the left;

f is negative at points arbitrarily close to $\lambda(f)$ on the left.

This set of f has π_σ-probability 1, by (57e, f). Let ϕ be a bounded, continuous function on $(-\infty, \infty)$. Then $\phi \circ \lambda$ is bounded, measurable, and continuous π_σ-almost everywhere. So

$$\int_{\mathscr{X}} \phi \circ \lambda \circ S_{(n)}\, d\mu \to \int_{C[0,1]} \phi \circ \lambda\, d\pi_\sigma.$$

But

$$L_n \leqq \lambda \circ S_{(n)} \leqq L_n + \frac{1}{n}. \qquad \bigstar$$

Here are the preliminaries to the proof of Donsker's theorem. To state (125–127), let (S, ρ) be a metric space. If A and C are subsets of S, let

$$\rho(A, C) = \inf\{\rho(a, c) : a \in A \text{ and } c \in C\}.$$

If $A = \{a\}$, abbreviate

$$\rho(a, C) = \rho(\{a\}, C).$$

(125) Lemma. *Let C_0 and C_1 be subsets of S with $\rho(C_0, C_1) > 0$. Then there is a ρ-uniformly continuous, real-valued function ϕ on S, with $0 \leqq \phi \leqq 1$, and $\phi = 0$ on C_0, and $\phi = 1$ on C_1.*

PROOF. Let $g(x) = \rho(x, C_0)$. Let

$$\gamma = \inf\{g(x) : x \in C_1\} = \rho(C_0, C_1) > 0.$$

You should check that $\phi = \min\{g/\gamma, 1\}$ works. ★

(126) Lemma. *Let φ be a bounded, real-valued function on S. Then there are bounded, real-valued, ρ-uniformly continuous functions ϕ_k and ψ_k on S, with*

(a) $\phi_k \leqq \phi_{k+1} \leqq \varphi \leqq \psi_{k+1} \leqq \psi_k$

(b) $\lim_{k\to\infty} \phi_k(x) = \varphi(x)$ *if φ is lower semicontinuous at x.*

(c) $\lim_{k\to\infty} \psi_k(x) = \varphi(x)$ *if φ is upper semicontinuous at x.*

As usual, φ is lower semicontinuous at x iff

$$\varphi(x) \leqq \liminf_{y\to x} \varphi(y).$$

And φ is upper semicontinuous at x iff

$$\varphi(x) \geqq \limsup_{y\to x} \varphi(y).$$

PROOF. Without real loss, suppose $0 < \varphi \leqq 1$. The discussion of ψ_k is omitted, being similar to that for ϕ_k. Using (125), construct a ρ-uniformly continuous, real-valued function $f(n, m, j)$ on S, bounded between 0 and j/n, such that:

$$f(n, m, j)(x) = 0 \qquad \text{for } x \text{ with } \varphi(x) \leqq j/n;$$

and

$$f(n, m, j)(x) = j/n \quad \text{for } x \text{ with } \rho(x, \{\varphi \leqq j/n\}) \geqq 1/m.$$

Clearly, $f(n, m, j) \leqq \varphi$. Let

$$\phi_k = \max\{f(n, m, j) : 1 \leqq n, m \leqq k \text{ and } 1 \leqq j \leqq n\}.$$

Clearly, $\phi_k \leqq \varphi$, and ϕ_k is nondecreasing with k, and ϕ_k is bounded as well as ρ-uniformly continuous. Suppose φ is lower semicontinuous at x. Let $j/n < \varphi(x) \leqq (j + 1)/n$. Then x is a positive ρ-distance from $\{\varphi \leqq j/n\}$, and $f(n, m, j)(x) = j/n$ for large m. ★

ASIDE. The argument shows $\lim \phi_k$ is the sup of all continuous $f \leqq \varphi$: that is, $\lim \phi_k$ is the largest semicontinuous $f \leqq \varphi$. Of course, $\lim \phi_k = f$ at x implies f is lower semicontinuous at x.

To state (127), let $\{P_n, P\}$ be probabilities on the σ-field generated by the ρ-open subsets of S.

(127) Lemma. *Suppose*

$$\int \phi \, dP_n \to \int \phi \, dP$$

for all bounded, ρ-uniformly continuous functions ϕ. Then

$$\int \varphi \, dP_n \to \int \varphi \, dP$$

for all bounded, measurable functions φ which are continuous P-almost everywhere.

NOTE ON MEASURABILITY. For a moment, let f be an arbitrary, real-valued function on S. Let F_n be the set of x for which there are y and z arbitrarily close to x, with

$$|f(y) - f(z)| > 1/n.$$

Then F_n is closed, and $\bigcup_n F_n$ is the set of x at which f is discontinuous. The continuity set of f is now revealed as a G_δ.

PROOF. Fix a bounded, measurable function φ which is continuous P-almost everywhere. Construct functions ϕ_k and ψ_k with properties (126a–c). Clearly,

$$\int \phi_k \, dP_n \leqq \int \varphi \, dP_n \leqq \int \psi_k \, dP_n.$$

So

$$\int \phi_k \, dP \leqq \liminf_n \int \varphi \, dP_n \leqq \limsup_n \int \varphi \, dP_n \leqq \int \psi_k \, dP. \tag{128}$$

But

$$P\{\lim_k \phi_k = \varphi = \lim_k \psi_k\} = 1.$$

Let $k \to \infty$ and use dominated convergence, to see that the extreme terms in (128) merge to $\int \varphi \, dP$. ★

To state (129), let Z_n and W_n be measurable maps from $(\Omega, \mathscr{F})$ to $(C[0, 1], \Sigma)$.

(129) Lemma. *Suppose*

$$\|Z_n - W_n\| \to 0 \quad \textit{in } \mathscr{P}\textit{-probability.}$$

Let φ be a bounded and uniformly continuous function on $C[0, 1]$. Then

$$\int \varphi(Z_n) \, d\mathscr{P} - \int \varphi(W_n) \, d\mathscr{P} \to 0.$$

PROOF. Let $\varepsilon > 0$. Let

$$A = \{\|Z_n - W_n\| > \varepsilon\}$$

$$\|\varphi\| = \sup \{|\varphi(f)| : f \in C[0, 1]\}$$

$$\operatorname{osc}(\varphi, \varepsilon) = \sup \{|\varphi(f) - \varphi(g)| : f, g \in C[0, 1] \text{ and } \|f - g\| \leqq \varepsilon\}.$$

Then

$$\left| \int \varphi(Z_n) \, d\mathscr{P} - \int \varphi(W_n) \, d\mathscr{P} \right| \leqq \int |\varphi(Z_n) - \varphi(W_n)| \, d\mathscr{P}$$

$$\leqq \eta_1 + \eta_2,$$

where:

$$\eta_1 = \int_A |\varphi(Z_n) - \varphi(W_n)| \, d\mathscr{P}$$

$$\leqq 2 \, \|\varphi\| \, \mathscr{P}\{A\};$$

and

$$\eta_2 = \int_{\Omega \setminus A} |\varphi(Z_n) - \varphi(W_n)| \, d\mathscr{P}$$

$$\leqq \operatorname{osc}(\varphi, \varepsilon).$$

Let $n \to \infty$: so $\mathscr{P}(A) \to 0$. Then let $\varepsilon \to 0$: so $\operatorname{osc}(\varphi, \varepsilon) \to 0$. ★

PROOF OF (120). Without real loss, suppose $\sigma^2 = 1$. Let $\tau_1, \tau_2, \ldots$ be the times in the Skorokhod representation (117) for F. Let

$$\hat{S}_n = B(\tau_n).$$

Define $\hat{S}_{(n)}$ in the usual way:

$$\hat{S}_{(n)}(\cdot, \omega) = \hat{S}_{(n)}(\omega) \in C[0, 1].$$

As (73a) implies, $\omega \to \hat{S}_{(n)}(\cdot, \omega)$ is a measurable map from $(\Omega, \mathscr{F})$ to $(C[0, 1], \Sigma)$. Let

$$B_n(t) = n^{-\frac{1}{2}}B(nt) \quad \text{for } 0 \leqq t \leqq 1.$$

As (72a) implies, $\omega \to B_n(\cdot, \omega)$ is a measurable map from $(\Omega, \mathscr{F})$ to $(C[0, 1], \Sigma)$: and so is

$$\omega \to \hat{S}_{(n)}(\cdot, \omega) - B_n(\cdot, \omega).$$

I claim

(130) $$\|\hat{S}_{(n)} - B_n\| \to 0 \quad \text{in } \mathscr{P}\text{-probability}.$$

To argue this, introduce Y_n:

$$Y_n(i/n) = B_n(i/n) \qquad \text{for } i = 0, \ldots, n$$

$$Y_n \quad \text{is linear on} \quad \left[\frac{i-1}{n}, \frac{i}{n}\right] \qquad \text{for } i = 1, \ldots, n.$$

Then (130) follows from

(131) $$\|Y_n - B_n\| \to 0 \quad \text{in } \mathscr{P}\text{-probability}$$

and

(132) $$\|\hat{S}_{(n)} - Y_n\| \to 0 \quad \text{in } \mathscr{P}\text{-probability}.$$

Argument for (131). Fix $\varepsilon > 0$. Then

$$\{\|Y_n - B_n\| > \varepsilon\} \subset \bigcup_{i=1}^n A_i,$$

where A_i is the event that

$$\tfrac{1}{2}\varepsilon n^{\frac{1}{2}} < \max\{|B(t) - B(i-1)| : i - 1 \leqq t \leqq i\}.$$

As (11c) implies,

$$\mathscr{P}\{A_i\} = \mathscr{P}\{A_1\} \quad \text{for } i = 1, \ldots, n.$$

As (16c) and (4a) imply:

$$\mathscr{P}\{A_1\} \leqq 4\mathscr{P}\{B(1) > \varepsilon n^{\frac{1}{2}}\} = 0[\exp(-\tfrac{1}{2}\varepsilon n^{\frac{1}{2}})].$$

So

$$\mathscr{P}\{\|Y_n - B_n\| > \varepsilon\} \leqq n\mathscr{P}\{A_1\} \to 0.$$

This completes the argument for (131).

Argument for (132). The style is like (118). Fix a positive ε and an r with $1 < r < 2$ and a positive integer M. Introduce the events

$$A = \{\|\hat{S}_{(n)} - Y_n\| > \varepsilon\}$$

$$C = \{\varepsilon n^{\frac{1}{2}} < \max [|B(\tau_m) - B(m)| : M \leqq m \leqq n]\}$$

$$D = \left\{r - 1 > \left|\frac{\tau_m}{m} - 1\right| \text{ for all } m \geqq M\right\}$$

$$E = \left\{\varepsilon n^{\frac{1}{2}} < \max \left[|B(t) - B(s)| : 0 \leqq s \leqq n \text{ and } s \leqq t \leqq \frac{r}{2 - r} s\right]\right\}$$

$$H = \{\varepsilon n^{\frac{1}{2}} < \max [|B(\tau_m) - B(m)| : 1 \leqq m \leqq M]\}.$$

By analysis,

$$A \subset C \cup H \quad \text{and} \quad C \cap D \subset E.$$

By logic,

$$C = (C \cap D) \cup (C \setminus D),$$

so

$$C \subset E \cup (\Omega \setminus D).$$

And

$$\mathscr{P}\{A\} \leqq \mathscr{P}\{E\} + \mathscr{P}\{\Omega \setminus D\} + \mathscr{P}\{H\}.$$

Fix a positive target error δ. By (19), there is an $r = r(\delta) > 1$ such that $\mathscr{P}\{E\} \leqq \delta/3$. By (117e), there is an $M = M(r, \delta)$ such that $\mathscr{P}\{\Omega \setminus D\} \leqq \delta/3$. By inspection, there is an $N = N(M, r, \delta) < \infty$ such that $n \geqq N$ implies $\mathscr{P}\{H\} \leqq \delta/3$. Then $\mathscr{P}\{A\} \leqq \delta$, completing the argument for (132), so for (130).

Let φ be bounded and uniformly continuous on $C[0, 1]$. By (129) and (130),

$$\int \varphi(\hat{S}_{(n)})\, d\mathscr{P} - \int \varphi(B_n)\, d\mathscr{P} \to 0.$$

By (117c, d), the μ-distribution of $\{S_n\}$ coincides with the $\mathscr{P}$-distribution of $\{\hat{S}_n\}$, so

$$\int_{\mathscr{X}} \varphi(S_{(n)})\, d\mu = \int_{\Omega} \varphi(\hat{S}_{(n)})\, d\mathscr{P}.$$

By scaling (11b), the $\mathscr{P}$-distribution of B_n coincides with the $\mathscr{P}$-distribution of B_1, namely π_1:

$$\int_{\Omega} \varphi(B_n)\, d\mathscr{P} = \int_{\Omega} \varphi(B_1)\, d\mathscr{P} = \int_{C[0,1]} \varphi\, d\pi_1.$$

This proves (120) for uniformly continuous φ. Now use (127), with the μ-distribution of $S_{(n)}$ for P_n, and π_1 for P. ★

8. STRASSEN'S INVARIANCE PRINCIPLE

Throughout this section, F is a distribution function on $(-\infty, \infty)$, with mean 0 and positive finite variance σ^2:

$$\int_{-\infty}^{\infty} x\, F(dx) = 0 \quad \text{and} \quad 0 < \int_{-\infty}^{\infty} x^2\, F(dx) = \sigma^2 < \infty.$$

Let $X_1, X_2, \ldots$ be independent random variables on some convenient probability triple $(\mathscr{X}, \mathscr{A}, \mu)$, with common distribution function F. Let $S_0 = 0$ and

$$S_n = X_1 + \cdots + X_n.$$

Remember $S_{(n)}$ from Section 7:

$$S_{(n)}\left(\frac{i}{n}, \omega\right) = n^{-\frac{1}{2}} S_i(\omega) \qquad \text{for } i = 0, \ldots, n$$

$$S_{(n)}(\cdot, \omega) \quad \text{is linear on} \quad \left[\frac{i-1}{n}, \frac{i}{n}\right] \qquad \text{for } i = 1, \ldots, n.$$

Let

$$T_n = (2 \log \log n)^{-\frac{1}{2}} S_{(n)}.$$

Remember that B is normalized Brownian motion on the triple $(\Omega, \mathscr{F}, \mathscr{P})$; convention (18) applies; and $(\Omega, \mathscr{F}, \mathscr{P})$ supports enough variables independent of B for (117) to hold. This artifact exists by (116). You should review the *preliminaries* and *compactness* subsections of Section 5. Remember that $C[0, 1]$ is the complete, separable, metric space of continuous functions on $[0, 1]$, with the sup metric d and the Borel σ-field Σ generated by the d-open sets.

Let K_σ be the set of absolutely continuous functions f on $[0, 1]$, with $f(0) = 0$ and

$$\int_0^1 f'(t)^2\, dt \leqq \sigma^2.$$

Suppose $\sigma > 0$. Then $f \in K_\sigma$ iff $\sigma^{-1} f \in K_1$. And K_1 was called K in Section 5.

Strassen's (1964) result is

(133) Theorem. *Let $\mathscr{X}_s$ be the set of $x \in \mathscr{X}$ such that: the indexed subset $\{T_n(x) : n \geqq 3\}$ of $C[0, 1]$ is relatively compact, with limit set K_σ. Then $\mathscr{X}_s \in \mathscr{A}$ and $\mu\{\mathscr{X}_s\} = 1$.*

PROOF. Without real loss, suppose $\sigma^2 = 1$. Let $\tau_1, \tau_2, \ldots$ be the times in the Skorokhod representation (117) for F. Let

$$\hat{S}_n = B(\tau_n) \quad \text{and} \quad \hat{T}_n = (2 \log \log n)^{-\frac{1}{2}}\, \hat{S}_{(n)}.$$

Let

$$Z_n(t) = (2n \log \log n)^{-\frac{1}{2}} B(nt) \quad \text{for } 0 \leqq t \leqq 1.$$

Following the pattern of (130), you should argue that

$$(134) \qquad \| \hat{T}_n - Z_n \| \to 0 \quad \text{with } \mathscr{P}\text{-probability } 1;$$

use estimate (20) in place of (19).

You can now use (134) and (81) to prove (133), with $\hat{T}$ replacing T. By (117c, d), the μ-distribution of $\{S_n\}$ coincides with the $\mathscr{P}$-distribution of $\{\hat{S}_n\}$. So the μ-distribution of $\{T_n\}$ coincides with the $\mathscr{P}$-distribution of $\{\hat{T}_n\}$. You can now use (74, 75) to replace $\hat{T}$ by T and get $\mathscr{X}_s \in \mathscr{A}$. ★

The next result is the iterated logarithm of Hartman and Wintner (1941).

(135) Corollary. *For μ-almost all $x \in \mathscr{X}$,*

$$\limsup_{n \to \infty} S_n(x)/(2n \log \log n)^{\frac{1}{2}} = \sigma^2$$

$$\liminf_{n \to \infty} S_n(x)/(2n \log \log n)^{\frac{1}{2}} = -\sigma^2.$$

PROOF. I will argue the first equality; the second is symmetric. Let

$$f(t) = \sigma t \quad \text{for } 0 \leqq t \leqq 1,$$

and let $x \in \mathscr{X}_s$. Then $f \in K_\sigma$ is in the limit set of $\{T_n(x)\}$; so for a subsequence $n' = n'(x)$,

$$T_{n'}(\cdot, x) \to f.$$

In particular,

$$T_{n'}(1, x) = (2n' \log \log n')^{-\frac{1}{2}} S_{n'}(x) \to f(1) = \sigma.$$

This proves:

$$\limsup_{n \to \infty} (2n \log \log n)^{-\frac{1}{2}} S_n(x) \geqq \sigma.$$

For the opposite inequality, keep $x \in \mathscr{X}_s$ and choose a subsequence $n' = n'(x)$ such that

$$(2n' \log \log n')^{-\frac{1}{2}} S_{n'}(x)$$

converges, say to L. By passing to a subsequence if necessary, suppose $T_{n'}(\cdot, x)$ converges, say to $f \in K_\sigma$. Then $f(1) = L$, and

$$\sigma \geqq \|f\| \geqq L,$$

using (79). ★

For example (136), fix $x \in \mathscr{X}_s$.

(136) Example. Let $n' = n'(x)$ be a subsequence with

$$(2n' \log \log n')^{-\frac{1}{2}} S_{n'}(x) \to \sigma.$$

Then $T_{n'}(\cdot, x) \to g$, where

$$g(t) = \sigma t \quad \text{for } 0 \leqq t \leqq 1.$$

PROOF. By passing to a subsequence if necessary, suppose $T_{n'}(\cdot, x)$ converges, say to $f \in K_\sigma$. Then

$$f(1) = \sigma = g(1).$$

Of course,

$$f(0) = 0 = g(0).$$

Using (80),

$$\sigma^2 \geqq \int_0^1 f'(t)^2 \, dt \geqq \int_0^1 g'(t)^2 \, dt = \sigma^2;$$

so (80) forces $f = g$. ★

Here is a preliminary (137) to examples (138–139). To state it, let H be a compact subset of the metric space M. Let

$$\{f_n : n \in N\}$$

be a relatively compact indexed subset of M, with limit set H. Let φ be a continuous function from M to the metric space S.

(137) Lemma. *The indexed subset $\{\varphi(f_n) : n \in N\}$ of S is relatively compact, with limit set $\varphi(H)$.*

PROOF. Easy. ★

For (138–139), confine x to $\mathscr{X}_s$; suppose $\sigma^2 = 1$; and write K for K_1.

(138) Example. The sequence of pairs $(T_n(\frac{1}{2}), T_n(1))$ is relatively compact, with limit set

$$L = \{(u, v) : u^2 + (v - u)^2 \leqq \tfrac{1}{2}\}.$$

Suppose $u^2 + (v - u)^2 = \frac{1}{2}$. Let n' be a random subsequence such that $T_{n'}(\frac{1}{2}) \to u$ and $T_{n'}(1) \to v$. Then $T_{n'} \to g$, where

$$\begin{aligned} g(t) &= 2ut && \text{for } 0 \leqq t \leqq \tfrac{1}{2} \\ &= u + 2(v - u)(t - \tfrac{1}{2}) && \text{for } \tfrac{1}{2} \leqq t \leqq 1. \end{aligned}$$

PROOF. By (137), the sequence of pairs $(T_n(\frac{1}{2}), T_n(1))$ is relatively compact, with limit set

$$\{(f(\tfrac{1}{2}), f(1)) : f \in K\}.$$

Now use (80) with $t_0 = 0, t_1 = \frac{1}{2}, t_2 = 1$ to see that $f \in K$ entails $(f(\frac{1}{2}), f(1)) \in L$. Conversely, if $(u, v) \in L$, and $g(0) = 0$, $g(\frac{1}{2}) = u$, $g(1) = v$, and g is linearly interpolated, then $g \in K$. If also $u^2 + (v - u)^2 = \frac{1}{2}$, then g is the only element of K taking the values u at $\frac{1}{2}$ and v at 1, by (80). ★

Here are some special cases of (138). To begin with,

$$\limsup_{n\to\infty} T_n(\tfrac{1}{2}) = 2^{-\frac{1}{2}} = \limsup_{n\to\infty} T_n(1) - T_n(\tfrac{1}{2}).$$

Along any subsequence n^* with $T_{n^*}(\frac{1}{2}) \to 2^{-1/2}$, however,

$$\limsup_{n\to\infty} T_{n^*}(1) - T_{n^*}(\tfrac{1}{2}) = 0.$$

If both $T_{n^*}(\frac{1}{2}) \to 2^{-1/2}$ and $T_{n^*}(1) - T_{n^*}(\frac{1}{2}) \to 0$, then $T_{n^*} \to g$, where

$$\begin{aligned} g(x) &= 2^{-1/2}x \quad \text{for } 0 \leqq x \leqq \tfrac{1}{2} \\ &= 2^{-1/2} \quad \text{for } \tfrac{1}{2} \leqq x \leqq 1. \end{aligned}$$

Along subsequences n^* with $T_{n^*}(1) \to 0$,

$$\limsup_n T_{n^*}(\tfrac{1}{2}) = \tfrac{1}{2}.$$

If both $T_{n^*}(1) \to 0$ and $T_{n^*}(\frac{1}{2}) \to \frac{1}{2}$, then $T_{n^*} \to g$, where

$$\begin{aligned} g(x) &= x \quad \text{for } 0 \leqq x \leqq \tfrac{1}{2} \\ &= 1 - x \quad \text{for } \tfrac{1}{2} \leqq x \leqq 1. \end{aligned}$$

★

For (139), let

$$L_n = (2n \log\log n)^{-\frac{1}{2}} \frac{1}{n} \Sigma_{i=1}^n S_i.$$

Keep x confined to $\mathscr{X}_s$, and $\sigma^2 = 1$, and $K = K_1$.

(139) Example. $\limsup_{n\to\infty} L_n = 3^{-\frac{1}{2}}$. Along a subsequence $n' = n'(x)$ for which $L_{n'} \to 3^{-\frac{1}{2}}$,

$$T_{n'}(\cdot, x) \to g;$$

where

$$g(t) = 3^{\frac{1}{2}}(t - \tfrac{1}{2}t^2) \quad \text{for } 0 \leqq t \leqq 1.$$

PROOF. Let $\varphi(f) = \int_0^1 f(t)\,dt$ for $f \in C[0, 1]$. Then φ is continuous. Suppose $f \in K$. By Fubini,

$$\begin{aligned} \varphi(f) &= \int_0^1 \int_0^t f'(s)\,ds\,dt \\ &= \int_0^1 \int_s^1 f'(s)\,dt\,ds \\ &= \int_0^1 (1 - s) f'(s)\,ds. \end{aligned}$$

By Schwarz,

$$\varphi(f)^2 \leqq \int_0^1 (1 - s)^2 \, ds \cdot \int_0^1 f'(s)^2 \, ds \leqq \tfrac{1}{3}.$$

So $\varphi(f) \leqq 3^{-\frac{1}{2}}$, and equality holds only if

$$f'(s) = c(1 - s) \quad \text{for } 0 \leqq s \leqq 1,$$

for some constant c. The largest c which keeps such an f in K is $c = 3^{1/2}$; then $f = g$ and $\varphi(f) = 3^{-\frac{1}{2}}$. If $c < 3^{\frac{1}{2}}$, then $\varphi(f) < 3^{-\frac{1}{2}}$. That is, φ retracted to K has a strict maximum of $3^{-\frac{1}{2}}$, assumed at g. In particular, (137) shows $\limsup \varphi(T_n) = 3^{-\frac{1}{2}}$. Suppose $\varphi(T_{n'}) \to 3^{-\frac{1}{2}}$. By passing to a subsequence if necessary, suppose $T_{n'} \to f \in K$. Then $\varphi(T_{n'}) \to \varphi(f) = 3^{-\frac{1}{2}} = \varphi(g)$, forcing $f = g$.

But what is the connection between L_n and $\varphi(T_n)$? Introduce

$$\tilde{T}_n(t, x) = (2n \log \log n)^{-\frac{1}{2}} \, S_i(x) \quad \text{for } \frac{i-1}{n} < t \leqq \frac{i}{n}$$

and $i = 1, \ldots, n$. Set $\tilde{T}_n(0, x) = 0$. For any bounded, measurable h, let

$$\|h\| = \sup \{|h(t)| : 0 \leqq t \leqq 1\} \quad \text{and} \quad \varphi(h) = \int_0^1 h(t) \, dt.$$

Then

$$L_n = \varphi(\tilde{T}_n).$$

Moreover,

$$\begin{aligned} |\varphi(\tilde{T}_n) - \varphi(T_n)| &\leqq \|\tilde{T}_n - T_n\| \\ &\leqq (2n \log \log n)^{-\frac{1}{2}} \max \{|X_i| : 1 \leqq i \leqq n\}. \end{aligned}$$

Even without the log log, this bound converges to 0 with μ-probability 1, by (*MC*, 3.9). ★

9. LINDEBERG'S THEOREM FOR MARTINGALES

The results of Sections 7 and 8 can be extended to cover martingales with a growth condition on the differences. Here, only the case of bounded differences will be considered, and only one result will be presented: an analog (143) of Lindeberg's theorem. The main part of this theorem is due to Lévy (1954a). Similar results have been obtained by Dvoretzky and by Strassen (1967). This section is somewhat apart from the rest of the book, and can be skipped. The argument is only sketched.

Let $X_1, X_2, \ldots$ be a sequence of random variables on the probability triple $(\mathscr{X}, \mathscr{A}, \mu)$. Let $\mathscr{A}_0, \mathscr{A}_1, \ldots$ be a nondecreasing sequence of sub σ-fields of $\mathscr{A}$. Let $\varepsilon > 0$. Then $(X_1, \mathscr{A}_0), (X_2, \mathscr{A}_1), \ldots$ is an *ε-system* provided:

(140a) $$X_n \text{ is } \mathscr{A}_n\text{-measurable};$$

(140b) $$|X_n| \leqq \varepsilon;$$

(140c) $$E(X_{n+1}|\mathscr{A}_n) = 0;$$

and

(140d) $$\Sigma_n v_n = \infty,$$

where

(141) $$v_{n+1} = E(X_{n+1}^2 | \mathscr{A}_n).$$

Define a process $\{S(t): 0 \leqq t \leqq 1\}$ in terms of such an ε-system, as follows. The process has continuous sample functions, $S(0) = 0$,

(142) $$S(v_1 + \cdots + v_n) = X_1 + \cdots + X_n,$$

and S is linearly interpolated. Of course, S can be defined in a natural way on $[0, \infty)$; but here S is to be considered only on $[0, 1]$. See Figure 9. The

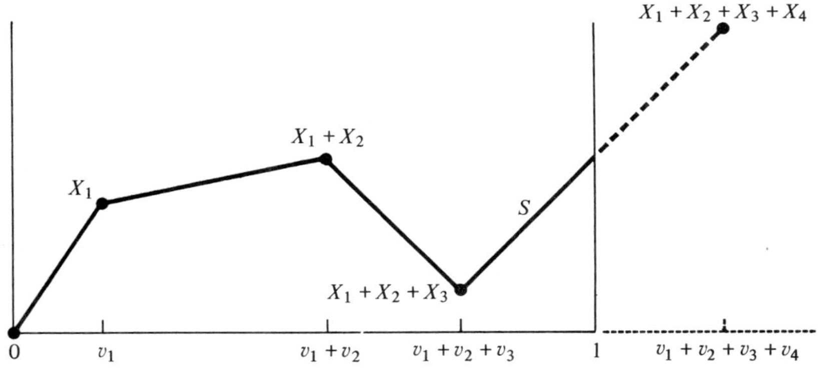

Figure 9.

main result of this section is that S, the trajectory of a martingale with uniformly small increments and standardized time scale, is approximately distributed like normalized Brownian motion on $[0, 1]$; this distribution was called π_1. To make this precise, fix a bounded, measurable function φ on $C[0, 1]$, which is continuous π_1-almost everywhere.

(143) Theorem. *There is a function $F(\varepsilon) = F(\varphi, \varepsilon)$ such that*

(a) $$\lim_{\varepsilon \to 0} F(\varepsilon) = 0$$

and

(b) $$\left| \int_{\mathscr{X}} \varphi(S)\, d\mu - \int_{C[0,1]} \varphi\, d\pi_1 \right| \leqq F(\varepsilon)$$

for all ε-systems.

Suppose $(X_1^*, \mathscr{A}_0^*), (X_2^*, \mathscr{A}_1^*), \ldots$ is another ε-system on another triple $(\mathscr{X}^*, \mathscr{A}^*, \mu^*)$, with conditional variances v_n^*, and standardized trajectory S^*. Suppose the distribution of $(X_1, X_2, \ldots; v_1, v_2, \ldots)$ coincides with that of $(X_1^*, X_2^*, \ldots; v_1^*, v_2^*, \ldots)$. Then the two ε-systems are called *equivalent*, and the distribution of S coincides with that of S^*. It is therefore enough to prove (143b) for any convenient representative of each equivalence class.

Fix your mind on one ε-system $(X_1, \mathscr{A}_0), (X_2, \mathscr{A}_1), \ldots$. The first part of the argument is designed to produce a convenient equivalent ε-system, embedded in a suitable way in normalized Brownian motion.

Let $(\Omega, \mathscr{F}, \mathscr{P})$ support a sequence $(B, W_1, W_2, \ldots)$ such that:

(144a) B is normalized Brownian motion;

(144b) W_n is uniformly distributed on $[0, 1]$;

(144c) $B, W_1, W_2, \ldots$ are independent.

Now let $\mathscr{X}^* = \mathscr{X} \times \Omega$ and $\mathscr{A}^* = \mathscr{A} \times \mathscr{F}$ and $\mu^* = \mu \times \mathscr{P}$. Further, let $\mathscr{A}_n^* = \mathscr{A}_n \times \Omega$ and $X_n^*(x, \omega) = X_n(x)$ and $B^*(x, \omega) = B(\omega)$ and $W_n^*(x, \omega) = W_n(\omega)$. Obviously, $(X_1^*, \mathscr{A}_0^*), \ldots$ is equivalent to the ε-system $(X_1, \mathscr{A}_0), \ldots$ and the conditional variances are given by $v_n^*(x, \omega) = v_n(x)$. Moreover, $(B^*, W_1^*, \ldots)$ is independent of $\mathscr{A}^*$, and in addition satisfies (144) with stars.

The next objective is to produce a sequence of random times $0 = \tau_0 \leqq \tau_1 \leqq \cdots$ on $\mathscr{X}^*$, with the property (145). To state (145), let $\mathscr{A}_n^+$ be the σ-field generated by $\mathscr{A}_n^*$ and by the pre-τ_n sigma field of B^* and by $W_1^*, \ldots, W_n^*$. Let $X_{n+1}^+ = B^*(\tau_{n+1}) - B^*(\tau_n)$.

(145a) $(X_1^+, \mathscr{A}_0^+), \ldots$ is equivalent to the ε-system $(X_1, \mathscr{A}_0), \ldots$

(145b) $$E(X_{n+1}^{+2} | \mathscr{A}_n^+) = v_{n+1}^*$$

(145c) $$E(\tau_{n+1} - \tau_n | \mathscr{A}_n^+) = v_{n+1}^*$$

(145d) $$\operatorname{var}(\tau_{n+1} - \tau_n | \mathscr{A}_n^+) \leqq 2\varepsilon^2 v_{n+1}^*.$$

(145e) τ_{n+1} is $\mathscr{A}_{n+1}^+$-measurable.

NOTE. The conditional variances of the $^+$-system coincide with those of the $*$-system.

This is done by the method of (117). Recall that $G(u, v)$ is the mean 0 distribution concentrated on $\{-u, v\}$. Use (108) to construct functions U_{n+1} and V_{n+1} on $\mathscr{X}^*$ which are measurable on the σ-field generated by $\mathscr{A}_n^*$ and W_{n+1}, such that $U_{n+1} \geqq 0$ and $V_{n+1} \geqq 0$ with $U_{n+1} = 0$ iff $V_{n+1} = 0$, and

$$E[G(U_{n+1}, V_{n+1})|\mathscr{A}_n^*]$$

is the conditional distribution of X_{n+1}^* given $\mathscr{A}_n^*$. Let $\tau_0 = 0$ and suppose τ_n defined. Then let τ_{n+1} be the least $t \geqq \tau_n$ with

$$B^*(t) - B^*(\tau_n) = -U_{n+1} \quad \text{or} \quad V_{n+1}.$$

Verify as in (117) that the conditional distribution of X_{n+1}^+ given $\mathscr{A}_n^+$ coincides with the conditional distribution of X_{n+1}^* given $\mathscr{A}_n^*$. In particular (145a, b) hold. The argument in (117) also proves (145c). Finally, (145d) is immediate from

(146) Lemma. *Let B be normalized Brownian motion. Let $a < 0 < b$ and let $\varepsilon = \max\{|a|, b\}$. Let τ be the least t with $B(t) = a$ or b. Then*

$$\operatorname{var} \tau \leqq E(\tau^2) \leqq 2\varepsilon^2 E(\tau).$$

PROOF. Verify that $Z(t) = B(t)^4 - 6B(t)^2 t + 3t^2$ is a martingale. As in (114), this entails $E[Z(\tau)] = 0$. So,

$$3E(\tau^2) \leqq 6E[B(\tau)^2 \tau] \leqq 6\varepsilon^2\, E(\tau),$$

because $|B(\tau)| \leqq \varepsilon$. ★

Theorem (143) now follows from the properties (145a–e) of this representation, as in (120). Let S^+ be the standardized trajectory of the $^+$-system. The main thing to prove is

(147) Proposition. *As $\varepsilon \to 0$, the distance $\max_{0 \leqq t \leqq 1} |S^+(t) - B^*(t)|$ converges to 0 in probability, uniformly in the original ε-system $(X_1, \mathscr{A}_0), \ldots$.*

PROOF. Let $n(t)$ be the least n with $v_1^* + \cdots + v_n^* \geqq t$. Let

$$u(t) = v_1^* + \cdots + v_{n(t)-1}^* \quad \text{and} \quad v(t) = v_1^* + \cdots + v_{n(t)}^*.$$

Thus, $t - \varepsilon^2 \leqq u(t) \leqq t \leqq v(t) \leqq t + \varepsilon^2$. Next, let

$$\begin{aligned} T_{n+1} &= \tau_{n+1} - \tau_n - v_{n+1}^* && \text{for } n + 1 \leqq n(1) \\ &= 0 && \text{for } n + 1 > n(1). \end{aligned}$$

Now

$$\{n + 1 = n(1)\} = \{v_1^* + \cdots + v_n^* < 1 \leqq v_1^* + \cdots + v_{n+1}^*\} \in \mathscr{A}_n^+,$$

so T_{n+1} is $\mathscr{A}_{n+1}^+$-measurable and its conditional expectation given $\mathscr{A}_n^+$ is 0 by (145c). Its conditional variance given $\mathscr{A}_n^+$ is 0 for $n + 1 > n(1)$, and at most $2\varepsilon^2 v_{n+1}^*$ for $n + 1 \leqq n(1)$ by (145d). Thus, the conditions of (148) below are satisfied with $a = 2\varepsilon^2(1 + \varepsilon^2)$. Fix $\delta > 0$. Let A be the set where

$$|\tau_n - (v_1^* + \cdots + v_n^*)| \leqq \delta$$

for all $n \leqq n(1)$. By (148), the complement of A has probability at most $2\varepsilon^2(1 + \varepsilon^2)/\delta^2$, uniformly in the original ε-system. Now

$$|\tau_{n(t)} - v(t)| \leqq \delta \quad \text{and} \quad |\tau_{n(t)-1} - u(t)| \leqq \delta$$

for $0 \leqq t \leqq 1$ on A. Thus

$$|\tau_{n(t)} - t| \leqq \delta + \varepsilon^2 \quad \text{and} \quad |\tau_{n(t)-1} - t| \leqq \delta + \varepsilon^2$$

for $0 \leqq t \leqq 1$ on A. Fix $\eta > 0$. Let C be the set where $0 \leqq s \leqq 2$ and $0 \leqq t \leqq 2$ and $|s - t| \leqq \delta + \varepsilon^2$ entail $|B^*(s) - B^*(t)| \leqq \eta$. As ε and δ tend to 0, the probability of the complement of C tends to 0, again without reference to the original ε-system. An explicit estimate is contained in (6). But $S^+(t)$ is between

$$X_1^+ + \cdots + X_{n(t)}^+ = B^*[\tau_{n(t)}] \quad \text{and} \quad X_1^+ + \cdots + X_{n(t)-1}^+ = B^*[\tau_{n(t)-1}].$$

Therefore, $|S^+(t) - B^*(t)| \leqq \eta$ for $0 \leqq t \leqq 1$ on $A \cap C$, proving (147); a convenient choice for δ is $\sqrt{\varepsilon}$. ★

(148) Lemma. *Let T_n be an $\mathscr{A}_n^+$-measurable function on $\mathscr{X}^*$. Let a and b be positive real numbers. Suppose*

$$E(T_{n+1}|\mathscr{A}_n^+) = 0 \quad \textit{and} \quad \Sigma \operatorname{var}(T_{n+1}|\mathscr{A}_n^+) \leqq a.$$

Then

$$\max_n(T_1 + \cdots + T_n) \geqq b \quad \textit{with probability at most } a/b^2.$$

PROOF. Check that $\{T_1 + \cdots + T_n : n = 1, 2, \ldots\}$ is a martingale with

$$E\{(T_1 + \cdots + T_n)^2\} \leqq a.$$

Then use Kolmogorov's inequality (Doob, 1953, Theorem 3.2 on p. 314). ★

EXAMPLE. Fix integers $0 = N_0 < N_1 < N_2 < \cdots$. Construct a process $\{S_n\}$ which is a symmetric Bernoulli walk for $N_{2k} \leqq n < N_{2k+1}$, and a symmetric Bernouilli walk with an absorbing barrier at 0 for those n with $N_{2k+1} \leqq n \leqq N_{2k+2}$. In more detail, let $S_0 = 0$ and let $X_n = S_n - S_{n-1}$ for $n = 1, 2, \ldots$. If

$$N_{2k} \leqq n < N_{2k+1}$$

or if

$$N_{2k+1} \leqq n < N_{2k+2} \quad \text{and} \quad S_n \neq 0,$$

then the conditional distribution of X_{n+1} given $X_1, \ldots, X_n$ is mass $\frac{1}{2}$ at each of ± 1. If

$$N_{2k+1} \leqq n < N_{2k+2} \quad \text{and} \quad S_n = 0,$$

then $X_{n+1} = 0$.

Let τ_m be the least n such that

$$\Sigma_{j=0}^{n} \operatorname{var}(X_{j+1}|X_1, \ldots, X_j) \geqq m.$$

Then $m^{-\frac{1}{2}}S_{\tau_m}$ is approximately $N(0, 1)$. You can use (143) on the process $\{m^{-\frac{1}{2}}X_j : j = 1, 2, \ldots\}$, which is an $m^{-\frac{1}{2}}$-system. Or you can check that $\{S_{\tau_m} : m = 0, 1, \ldots\}$ is a perfectly ordinary symmetric Bernoulli walk, as Dick Drogin suggests. If $N_{2k+2} - N_{2k+1}$ is large by comparison with N_{2k+1}, then

$$\text{Prob}\,\{S_{N_{2k}} = 0\} \to 1 \quad \text{as } k \to \infty;$$

so there are no factors f_n, constant or random, which make

$$\text{Distribution } f_n S_n \to N(0, 1).$$

★

10. CENTRAL LIMIT THEOREM FOR RANDOM SUMS

Let $X_1, X_2, \ldots$ be independent and identically distributed, having mean 0 and variance 1, on the triple $(\Omega, \mathscr{F}, \mathscr{P})$. Let

$$S(n) = X_1 + \cdots + X_n.$$

Let $\tau_1, \tau_2, \ldots$ be a sequence of positive integer-valued random variables, such that

(149a) $$\frac{\tau_n}{n} \to \tau \quad \text{in } \mathscr{P}\text{-probability},$$

where

(149b) $$\tau \quad \text{is positive and finite.}$$

Do not assume that the τ_n are Markov times.

Let $\Phi(v)$ be the normal distribution with mean 0 and variance v. The object of this section is to prove the following result (Wittenberg, 1964).

(150) Theorem. *Suppose* (149). *The distribution of* $S(\tau_n)/n^{\frac{1}{2}}$ *weak* converges to*

$$\int \Phi(\tau)\, d\mathscr{P}.$$

PROOF. Let $\Phi(v, f)$ be the $\Phi(v)$-expectation of f, for bounded, measurable f. Then

$$v \to \Phi(v, f)$$

is continuous. It is enough to prove

$$\int f[S(\tau_n)/n^{\frac{1}{2}}]\,d\mathscr{P} \to \int \Phi(\tau, f)\,d\mathscr{P}$$

for all bounded, continuous f with a bounded derivative f'. Let

$$\|g\| = \sup\{|g(x)|\},$$

so

$$|f(x) - f(y)| \leqq \|f'\| \cdot |x - y|.$$

Let m be an integer and $r > 1$. Interpret $S(nr^m)$ as $S(j)$, where j is the closest positive integer to nr^m. As $n \to \infty$ with m and r fixed, the distribution of $S(nr^m)/n^{\frac{1}{2}}$ tends to $\Phi(r^m)$ by (121). Furthermore, the tail σ-field of $\{S(n)\}$ is trivial by the Hewitt-Savage 0–1 Law (*MC*, 1.22). Use (*MC*, 1.129) to conclude

(151) $$\lim_{n\to\infty} \int_F f[S(nr^m)/n^{\frac{1}{2}}]\,d\mathscr{P} = \Phi(r^m, f) \cdot \mathscr{P}\{F\}$$

for any $F \in \mathscr{F}$.

Fix $\delta > 0$. Use (149) to choose finite $a > 1$ so that $\mathscr{P}\{A_n\} \geqq 1 - \delta$ for all n, where

$$A_n = A \cap \left\{\frac{1}{a} \leqq \frac{\tau_n}{n} < a\right\}$$

and

$$A = \left\{\frac{1}{a} \leqq \tau < a\right\}.$$

Fix $\varepsilon > 0$. Use (118) to choose positive integers M and N so large that $\mathscr{P}\{C_n\} \geqq 1 - \delta$ for all $n \geqq N$, where C_n is the event

$$\varepsilon n^{\frac{1}{2}} \geqq \max\{|S(j) - S(k)| : 0 \leqq j \leqq arn \text{ and } j \leqq k \leqq r^3 j\}$$

and

$$r = a^{1/M}.$$

Use (149) again to choose N' with $N \leqq N' < \infty$ so that $\mathscr{P}\{D_n\} \geqq 1 - \delta$ for $n \geqq N'$, where

$$D_n = \left\{\frac{1}{r} \leqq \frac{\tau_n}{n\tau} \leqq r\right\}.$$

Let $G_n = A_n \cap C_n \cap D_n$, so

(152) $$\mathscr{P}\{G_n\} \geqq 1 - 3\delta \quad \text{for } n \geqq N'.$$

For $n \geqq N'$,

(153)
$$\left| \int f[S(\tau_n)/n^{\frac{1}{2}}]\, d\mathscr{P} - \int_{G_n} f[S(\tau_n)/n^{\frac{1}{2}}]\, d\mathscr{P} \right| \leqq \|f\| \cdot [1 - \mathscr{P}\{G_n\}] \leqq 3\|f\|\delta.$$

Let $H_m = \{r^m \leqq \tau < r^{m+1}\}$, and

$$S_M = \Sigma_{m=-M}^{M-1} \int_{G_n \cap H_m} f[S(nr^m)/n^{\frac{1}{2}}]\, d\mathscr{P}.$$

I claim

(154)
$$\left| \int_{G_n} f[S(\tau_n)/n^{\frac{1}{2}}]\, d\mathscr{P} - S_M \right| \leqq \|f'\|\varepsilon.$$

Indeed, $a^{-1} \leqq \tau < a$ on $A \subset G_n$ and $r = a^{1/M}$; so

$$\int_{G_n} f[S(\tau_n)/n^{\frac{1}{2}}]\, d\mathscr{P} = \Sigma_{m=-M}^{M-1} \int_{G_n \cap H_m} f[S(\tau_n)/n^{\frac{1}{2}}]\, d\mathscr{P}.$$

Keep m between $-M$ and $M - 1$. On $D_n \cap H_m$,

$$nr^{m-1} \leqq \tau_n < nr^{m+2}.$$

On $C_n \cap D_n \cap H_m$,

$$|S(\tau_n)/n^{\frac{1}{2}} - S(nr^m)/n^{\frac{1}{2}}| \leqq \varepsilon;$$

so

$$|f[S(\tau_n)/n^{\frac{1}{2}}] - f[S(nr^m)/n^{\frac{1}{2}}]| \leqq \|f'\|\varepsilon.$$

Therefore,

$$\left| \int_{G_n \cap H_m} f[S(\tau_n)/n^{\frac{1}{2}}]\, d\mathscr{P} - \int_{G_n \cap H_m} f[S(nr^m)/n^{\frac{1}{2}}]\, d\mathscr{P} \right| \leqq \|f'\|\mathscr{P}\{G_n \cap H_m\}\varepsilon.$$

The sum of this error from $m = -M$ to $M - 1$ is at most $\|f'\|\varepsilon$, proving (154).

Let

$$T_M = \Sigma_{m=-M}^{M-1} \int_{H_m} f[S(nr^m)/n^{\frac{1}{2}}]\, d\mathscr{P}.$$

I claim that for $n \geqq N'$,

(155)
$$|S_M - T_M| \leqq 3\|f\|\delta.$$

Indeed,

$$\left|\int_{G_n\cap H_m} f[S(nr^m)/n^{\frac{1}{2}}]\,d\mathscr{P} - \int_{H_m} f[S(nr^m)/n^{\frac{1}{2}}]\,d\mathscr{P}\right| \leqq \|f\|\cdot\mathscr{P}\{H_m \cap (\Omega\backslash G_n)\}.$$

The sum of this error from $m = -M$ to $M-1$ is at most $3\|f\|\delta$ by (152), proving (155).

Add (153–155): for $n \geqq N'$,

$$\left|\int f[S(\tau_n)/n^{\frac{1}{2}}]\,d\mathscr{P} - T_M\right| \leqq 6\|f\|\delta + \|f'\|\varepsilon.$$

As $n \to \infty$, the sum T_M tends by (151) to

$$U_M = \Sigma_{m=-M}^{M+1}\,\Phi(r^m, f)\cdot\mathscr{P}\{H_m\}.$$

Keep a fixed, and let $M \uparrow \infty$, so $r = a^{1/M} \downarrow 1$. Then $N' \to \infty$, but this is harmless. The sum U_M converges to

$$V_a = \int_A \Phi(\tau, f)\,d\mathscr{P}.$$

Let $a \uparrow \infty$. The integral V_a tends to

$$\int \Phi(\tau, f)\,d\mathscr{P}.$$

★

(156) Corollary. *Suppose* (149). *The distribution of* $S(\tau_n)/\tau_n^{\frac{1}{2}}$ *tends to* $N(0, 1)$.

PROOF. The argument for the theorem shows

$$\lim_{n\to\infty}\int_F f[S(\tau_n)/n^{\frac{1}{2}}]\,d\mathscr{P} = \int_F \Phi(\tau, f)\,d\mathscr{P}$$

for $F \in \mathscr{F}$ and bounded, continuous f with bounded derivative f'. Specialize $F = \{\tau \leqq x\}$, to see that the joint distribution of τ and $S(\tau_n)/n^{\frac{1}{2}}$ converges to the joint distribution of T and S, where: T is distributed like τ, and given $T = t$, the conditional distribution of S is $\Phi(t)$. Consequently, the joint distribution of τ_n/n and $S(\tau_n)/n^{\frac{1}{2}}$ converges to the joint distribution of T and S. Finally, the distribution of

$$S(\tau_n)/\tau_n^{\frac{1}{2}} = (\tau_n/n)^{-\frac{1}{2}}\,S(\tau_n)/n^{\frac{1}{2}}$$

converges to the distribution of $T^{-\frac{1}{2}}S$, namely $N(0, 1)$. ★

The next example shows that condition (149a) cannot be weakened to

τ_n/n converges in distribution.

(157) Example. Let $\tau_n = n$ when $S_n > 0$, and $\tau_n = 2n$ when $S_n \leqq 0$. Then

$$\mathscr{P}\{\tau_n/n = 1\} \to \tfrac{1}{2}.$$

But the distribution of $S(\tau_n)/n^{\frac{1}{2}}$ does not tend to $\frac{1}{2}\Phi(1) + \frac{1}{2}\Phi(2)$. Indeed,

$$\mathscr{P}\{S(\tau_n) > 0\} = \mathscr{P}\{S(n) > 0\} + \mathscr{P}\{S(n) \leqq 0 < S(2n)\}$$
$$\to \mathscr{P}\{Y > 0\} + \mathscr{P}\{Y \leqq 0 < Y + Z\},$$

where Y and Z are independent $N(0, 1)$ variables. This limit is more than $\frac{1}{2}$. Both limiting operations can be justified by (120). ★

11. WALD'S IDENTITY

My object in this section is to present a version of Wald's identity that I learned from Larry Shepp. Remember that $C[0, \infty)$ is the space of continuous, real-valued functions on $[0, \infty)$, and ξ is the coordinate process:

$$\xi(s, f) = f(s) \quad \text{for } 0 \leqq s < \infty \text{ and } f \in C[0, \infty).$$

Give $C[0, \infty)$ the σ-field Σ generated by ξ, and let $\Sigma(t)$ be the sub-σ-field generated by $\xi(s)$ with $0 \leqq s \leqq t$. Let τ be a stopping time on $C[0, \infty)$: that is, τ is nonnegative and

$$\{\tau < t\} \in \Sigma(t) \quad \text{for all } t > 0.$$

Let λ be a real number, and let

$$B_\lambda(t) = B(t) + \lambda t;$$

as usual, B is standard Brownian motion. So B_λ is Brownian motion with drift λ.

Wald's identity is

$$\text{(158)} \qquad \int_{\{\tau(B) < \infty\}} e^{\lambda B[\tau(B)] - \frac{1}{2}\lambda^2 \tau} \, d\mathscr{P} = \mathscr{P}\{\tau(B_\lambda) < \infty\}.$$

Somewhat more is true. Let Q_λ be the distribution of B_λ, so Q_λ is a probability on $(C[0, \infty), \Sigma)$. Let $\Sigma(\tau+)$ be the σ-field of all $A \in \Sigma$ such that

$$A \cap \{\tau < t\} \in \Sigma(t) \quad \text{for } t > 0.$$

(159) Theorem. *Q_λ is absolutely continuous with respect to Q_0 when both are retracted to $\{\tau < \infty\} \cap \Sigma(\tau+)$; then a Radon–Nikodym derivative of Q_λ with respect to Q_0 is*

$$e^{\lambda \xi(\tau) - \frac{1}{2}\lambda^2 \tau}.$$

DISCUSSION. The process

$$t \to e^{\lambda B(t) - \frac{1}{2}\lambda^2 t}$$

is a martingale. For each t, expand and collect terms: let $C_n(t)$ be the coefficient of λ^n, so $C_0(t) = 1$. Then C_n is a martingale with mean 0, for $n \geqq 1$.

Suppose τ is a stopping time on $C[0, \infty)$, with

$$Q_\lambda\{\tau < \infty\} = 1 \quad \text{for all } \lambda.$$

Let $\sigma = \tau(B)$. Then

$$E\{C_n(\sigma)\} = 0 \quad \text{for } n \geqq 1.$$

This accounts for (114) and (146).

PROOF FOR CONSTANT τ. Let $0 < s < \infty$. Then

$$\frac{1}{\sqrt{2\pi s}} \int_{-\infty}^{x} e^{-[(u-\lambda s)^2/2s]}\, du = \frac{1}{\sqrt{2\pi s}} \int_{-\infty}^{x} e^{\lambda u - \frac{1}{2}\lambda^2 s}\, e^{-u^2/2s}\, du.$$

That is,

$$(160) \qquad Q_\lambda\{\xi(s) \leqq x\} = \int_{\{\xi(s) \leqq x\}} e^{\lambda\xi(s) - \frac{1}{2}\lambda^2 s}\, dQ_0.$$

Let $0 = t_0 < t_1 < \cdots < t_n = t$. Put $s_m = t_m - t_{m-1}$ and $\zeta_m = \xi(t_m) - \xi(t_{m-1})$. Let $x_0, x_1, \ldots, x_n$ be real numbers. Let

$$D = \{\zeta_m \leqq x_m \text{ for } m = 1, \ldots, n\}.$$

Clearly,

$$(161) \qquad t = \Sigma_{m=1}^n s_m \quad \text{and} \quad \xi(t) = \Sigma_{m=1}^n \zeta_m.$$

With respect to Q_λ or Q_0,

$$(162) \qquad \zeta_m \text{ is distributed like } \xi(s_m),$$

and

$$(163) \qquad \text{the } \zeta_m \text{ are independent.}$$

Now compute.

$$\begin{aligned}
Q_\lambda\{D\} &= \Pi_{m=1}^n Q_\lambda\{\zeta_m \leqq x_m\} && \text{by (163)}\\
&= \Pi_{m=1}^n \int_{\{\zeta_m \leqq x_m\}} e^{\lambda\zeta_m - \frac{1}{2}\lambda^2 s_m}\, dQ_0 && \text{by (160, 162)}\\
&= \int_D \Pi_{m=1}^n e^{\lambda\zeta_m - \frac{1}{2}\lambda^2 s_m}\, dQ_0 && \text{by (163)}\\
&= \int_D e^{\lambda\xi(t) - \frac{1}{2}\lambda^2 t}\, dQ_0 && \text{by (161).}
\end{aligned}$$

As the t_m and x_m and n vary, D spans Σ. ★

Using (16), check

(164) $$\sup_{0 \leqq s \leqq t} e^{\lambda B(s) - \frac{1}{2}\lambda^2 s} \in L^1(Q_0).$$

The rest of the argument is best carried on in a general setting. Let $(\mathscr{X}, \Sigma)$ be a measurable space. For $0 \leqq t < \infty$, let $\Sigma(t)$ be a sub-σ-field of Σ; suppose $\Sigma(t)$ is nondecreasing with t. Let τ be a stopping time: that is, τ is nonnegative and

$$\{\tau < t\} \in \Sigma(t) \quad \text{for all } t > 0.$$

Let $\Sigma(\tau+)$ be the σ-field of $A \in \Sigma$ such that

$$A \cap \{\tau < t\} \in \Sigma(t) \quad \text{for all } t > 0.$$

Let P and Q be two probabilities on $(\mathscr{X}, \Sigma)$. Suppose P is absolutely continuous with respect to Q on each $\Sigma(t)$, with Radon–Nikodym derivative $Y(t)$. Suppose Y can be chosen so that

(165) $Y(\cdot, \omega)$ is right continuous

(166) $$\sup_{0 \leqq s \leqq t} Y(s) \in L^1(Q) \quad \text{for all } t > 0.$$

(167) Proposition. *P is absolutely continuous with respect to Q when both are retracted to $\{\tau < \infty\} \cap \Sigma(\tau+)$; then a Randon–Nikodym derivative of P with respect to Q is $Y(\tau)$.*

PROOF. Let $A \in \Sigma(\tau+)$ and $A \subset \{\tau < \infty\}$. I have to prove

$$P\{A\} = \int_A Y(\tau)\, dQ,$$

which follows from

$$P\{A \text{ and } \tau < t\} = \int_{\{A \text{ and } \tau < t\}} Y(\tau)\, dQ$$

for all t. Without real loss, put $t = 1$. Let τ_n be the least $j/2^n > \tau$. Now

$$\{A \text{ and } \tau_n = j/2^n\} \in \Sigma(j/2^n),$$

so

$$P\{A \text{ and } \tau_n = j/2^n\} = \int_{\{A \text{ and } \tau_n = j/2^n\}} Y(j/2^n)\, dQ$$
$$= \int_{\{A \text{ and } \tau_n = j/2^n\}} Y(\tau_n)\, dQ.$$

Sum out $j = 1, \ldots, 2^n$:

$$P\{A \text{ and } \tau < 1\} = \int_{\{A \text{ and } \tau < 1\}} Y(\tau_n)\, dQ.$$

But $Y(\tau_n) \to Y(\tau)$ by (165), and (166) bounds this sequence above by an L^1 function, on $\{\tau < 1\}$. So dominated convergence wins again. ★

DISCUSSION. The only real condition is (166): everything else is normalization. And (166) is roughly the same as

$$Y(t) \log [1 + Y(t)] \in L^1(Q) \quad \text{for all } t > 0.$$

See (Blackwell and Dubins, 1963) and (Doob, 1953, Theorem 3.4 on p. 317). You may also want to look at (3.29).

2

DIFFUSION

1. INTRODUCTION

The material in this chapter is based on (Itô and McKean, 1965). I also benefitted from reading the diffusion chapter in (Breiman, 1968), and referred on occasion to (Blumenthal and Getoor, 1968) and to (Dynkin, 1965). There are few further bibliography citations in this chapter, because I have no real acquaintance with the relevant literature.

Let I be an interval of real numbers; for example, $(-\infty, \infty)$ or $(0, 1]$. Let $\Omega = \Omega_I$ be the set of all continuous functions from $[0, \infty)$ to I. Let X be the coordinate process on Ω:

$$X(t, \omega) = \omega(t) \quad \text{for } 0 \leqq t < \infty \quad \text{and} \quad \omega \in \Omega.$$

Endow Ω with the Borel σ-field $\mathscr{B}$ generated by $\{X(t):0 \leqq t < \infty\}$. Let $\mathscr{B}(t)$ be the σ-field generated by $\{X(s):0 \leqq s \leqq t\}$, and let

$$\mathscr{B}(t+) = \bigcap_{\varepsilon>0} \mathscr{B}(t + \varepsilon).$$

Let τ be a random variable on Ω with values in $[0, \infty]$. Then τ is a *Markov time* iff $\{\tau < t\} \in \mathscr{B}(t)$ for all $t \in [0, \infty)$. Let $\mathscr{B}(\tau+)$ be the σ-field of all $A \in \mathscr{B}$ such that for all $t \in [0, \infty)$,

$$A \cap \{\tau < t\} \in \mathscr{B}(t).$$

The two definitions of $\mathscr{B}(t+)$ are consistent.

Let S_t be the shift mapping of Ω into itself: $S_t(\omega) = \omega(t + \cdot)$. As usual, S_τ maps Ω into itself as follows:

$$(S_\tau)(\omega) = S_{\tau(\omega)}(\omega);$$

I would like to thank Mrs. S. T. Moy, who kept me reasonably honest the first time I taught diffusion; and Howard Taylor, who checked the final draft of this chapter.

or more explicitly,

$$[(S_\tau)(\omega)](s) = \omega[\tau(\omega) + s].$$

For any measurable τ, the mapping S_τ is measurable.

For each $x \in I$, let P_x be a probability on Ω. Then $\{P_x\}$ is a *diffusion* iff $P_x\{X(0) = x\} = 1$ and $\{P_x\}$ has the strong Markov property. Formally,

(1) Definition. *$\{P_x : x \in I\}$ is a* diffusion *on I iff*

(a) *$P_x\{X(0) = x\} = 1$ for all $x \in I$, and*

(b) *for all x and all Markov times τ, the conditional P_x-distribution of S_τ given $\mathscr{B}(\tau+)$ is $P_{X(\tau)}$ on $\{\tau < \infty\}$.*

Here, $P_{X(\tau)}$ is the function

$$(\omega, A) \to P_{X[\tau(\omega),\omega]}\{A\}$$

of pairs (ω, A) with $\omega \in \{\tau < \infty\}$ and A a measurable subset of Ω. In particular, P_x is Markov with stationary transitions, starting from x, as defined on page 31. Measurability of $x \to P_x\{x\}$ is assumed.

Property (1b) can be restated as follows. For $x \in I$, Markov times τ, sets $A \in \mathscr{B}(\tau+)$ with $A \subset \{\tau < \infty\}$,

$$P_x\{A \text{ and } S_\tau \in B\} = \int_A P_{X[\tau(\omega),\omega]}\{B\}\, P_x(d\omega). \tag{1c}$$

A trifle more generally, for bounded $\mathscr{B}(\tau+)$-measurable functions ϕ on $C(I)$ which vanish on $\{\tau = \infty\}$, and bounded $\mathscr{B}$-measurable ψ,

$$\int \phi \cdot \psi(S_\tau)\, dP_x = \int \phi(\omega) \cdot \int \psi(\omega')\, P_{X[\tau(\omega),\omega]}(d\omega')\, P_x(d\omega). \tag{1d}$$

Let $\{B(t) : 0 \leqq t < \infty\}$ be normalized Brownian motion on the triple $(\mathscr{X}, \Sigma, \mu)$; see definition (1.1). For any real x, let B_x be the μ-distribution of $x + B$. Then $\{B_x\}$ is *standard Brownian motion*. For any x and $\sigma > 0$, let σB_x be the μ-distribution of $x + \sigma B(\cdot)$. Then $\{\sigma B_x\}$ is *standard Brownian motion with variance parameter σ^2*.

(2) Theorem. *Standard Brownian motion is a diffusion on $(-\infty, \infty)$.*

PROOF. Use (1.24). ★

Let F be continuous and strictly monotone, increasing or decreasing, on I. Let Q_x be the $P_{F^{-1}(x)}$-distribution of $\{F[X(t)] : 0 \leqq t < \infty\}$ for $x \in F(I)$.

(3) Theorem. *If $\{P_x\}$ is a diffusion on I, then $\{Q_x\}$ is a diffusion on $F(I)$.*

PROOF. Easy. ★

I will now state the main results of this chapter under some simplifying assumptions. Suppose

$$\{P_x\} \text{ is a diffusion on } I = (-\infty, \infty)$$

and

$$P_x\{\text{hit } y\} = 1 \quad \text{for all } x \text{ and } y \text{ in } I.$$

As usual

$$\{\text{hit } y\} = \{\omega : \omega \in \Omega \text{ and } \omega(t) = y \quad \text{for some } t \in [0, \infty)\}$$

$$= \bigcup_{n=1}^{\infty} \{\omega : \omega \in \Omega \text{ and } \inf_r |\omega(r) - y| = 0 \text{ for rational } r \text{ with } 0 \leqq r \leqq n\}$$

is measurable. Abbreviate $S(\pm\infty) = \lim_{x \to \pm\infty} S(x)$.

(A) Theorem. *There is a continuous, strictly increasing function S, with* $S(-\infty) = -\infty$ *and* $S(\infty) = \infty$, *such that for* $a \leqq x \leqq b$,

$$P_x\{\textit{hit } b \textit{ before } a\} = \frac{S(x) - S(a)}{S(b) - S(a)}.$$

This function S is the *scale function* of $\{P_x\}$; if S is the identity function, that is $S(x) = x$, then $\{P_x\}$ is in *natural scale*. Let Q_x be the $P_{S^{-1}(x)}$-distribution of $\{S[X(t)] : 0 \leqq t < \infty\}$.

(B) Fact. $\{Q_x\}$ *is a diffusion in natural scale on* $(-\infty, \infty)$.

Theorem (A) is proved in Section 3; Fact (B) is immediate from (3) and (A). For the rest of this section, suppose

$$\{P_x\} \quad \text{is in natural scale.}$$

This presents no loss of generality by (B). Let $a \leqq b$ be in I. Let τ_{ab} be the least t if any with $X(t) = a$ or b, and $\tau_{ab} = \infty$ if none. Write E_x for expectation with respect to P_x, and let $e_{ab}(x) = E_x(\tau_{ab})$ for $a \leqq x \leqq b$. Let f^+ be the right derivative of f. A measure m is *strictly positive* iff any nonempty open set has positive m-measure; and m is *locally finite* iff any compact set has finite m-measure.

(C) Theorem. $e_{ab}(\cdot)$ *is strictly concave on* $[a, b]$. *There is a unique measure* m *on* $(-\infty, \infty)$, *such that for* $a < \alpha \leqq \beta < b$,

$$e_{ab}^+(\alpha) - e_{ab}^+(\beta) = m(\alpha, \beta].$$

Moreover, m is strictly positive and locally finite.

This measure m is the *speed measure* of $\{P_x\}$. Theorem (C) is proved in Section 6. The speed measure of standard Brownian motion with variance parameter σ^2 is Lebesgue measure divided by $\sigma^2/2$. Let $BC(I)$ be the set of bounded, continuous real-valued functions on I. As usual,

$$\|f\| = \sup\{|f(x)| : x \in I\}.$$

For $\varepsilon \geqq 0$, and $f \in BC(I)$, let

$$(T_\varepsilon f)(x) = E_x\{f[X(\varepsilon)]\},$$

so T_ε maps $BC(I)$ into itself. Let $\Gamma_\varepsilon = \varepsilon^{-1}(T_\varepsilon - 1)$ and $\Gamma_0 = \lim_{\varepsilon \to 0} \Gamma_\varepsilon$ on the set Δ_0 of $f \in BC(I)$ such that $\Gamma_\varepsilon f$ converges pointwise and boundedly to an element of $BC(I)$. The *infinitesimal generator* of $\{P_x\}$ is (Γ_0, Δ_0). As will be seen in Section 4,

(D) Theorem. *$\{P_x\}$ is determined by its infinitesimal generator.*

Recall that $\{P_x\}$ is in natural scale. As will be seen in Section 7,

(E) Theorem. *Δ_0 is the set of $f \in BC(I)$ such that f^+ exists, is finite and continuous from the right, and of bounded variation on finite intervals, with*

$$df^+ = g\,dm$$

for some $g \in BC(I)$. This g is unique, and is $\Gamma_0 f$.

As is immediate from (D) and (E),

(F) Corollary. *$\{P_x\}$ is determined by its speed measure.*

To state the last result, let m be a locally finite measure on $(-\infty, \infty)$. Suppose m is absolutely continuous with respect to Lebesgue measure, and has a continuous positive density $\dot{m}$. Let

$$M(t) = \frac{1}{2}\int_0^t \dot{m}[X(s)]\,ds.$$

Then $M(0, \omega) = 0$; while $M(\cdot, \omega)$ is continuous and strictly increasing; and $M(\infty, \omega) = \infty$. Let $T(\cdot, \omega)$ be the inverse of $M(\cdot, \omega)$, so $T(0, \omega) = 0$; while $T(\cdot, \omega)$ is continuous and strictly increasing; and $T(\infty, \omega) = \infty$. Let $Y = X \circ T$, that is,

$$Y(t, \omega) = X[T(t, \omega), \omega].$$

Let Q_x be the B_x-distribution of Y, where $\{B_x\}$ is standard Brownian motion. As will be proved in Section 9,

(G) Theorem. *$\{Q_x\}$ is a diffusion on $(-\infty, \infty)$ in natural scale, with speed measure m.*

In particular, if m was the speed measure of $\{P_x\}$, then $P_x = Q_x$ for all x by (F). Informally, a process with distribution P_x can be obtained by changing the time scale of a Brownian motion process, as follows. Fix a Brownian path f, and suppose that t on the Brownian time scale corresponds to $t^* = t^*(f)$

on the diffusion time scale. Then $t + dt$ on the Brownian scale goes to $t^* + \frac{1}{2}\dot{m}[f(t)]\,dt$ on the diffusion scale. Another way to visualize the diffusion is this. Suppose the process is in the position y at time t, and its past is known. The position of the process at time $t + dt$ is distributed conditionally like a normal random variable, with mean y and variance $[2/\dot{m}(y)]\,dt$.

Theorem (G) can be generalized in two different directions: see (Blumenthal, Getoor, and McKean, 1962) and (Dubins and Schwarz, 1965). A similar result for countable state processes was presented in Section 2.5 of *ACM*. This situation is unsatisfactory, and there is probably one quite general theorem which includes all these cases.

On a first reading of this chapter, do only (92–93) in Section 8; in Section 9, suppose the speed measure is absolutely continuous, with a continuous density; skim Sections 10–15.

You may wish to read Section 6.6 of *MC*, on Markov times.

2. REGULARITY

Suppose $\{P_x\}$ is a diffusion on I.

(4) Lemma. (*Blumenthal* 0–1 *Law*). *The σ-field $\mathscr{B}(0+)$ is P_x-trivial.*

PROOF. Let $\tau \equiv 0$; so τ is Markov, S_τ is the identity, $\mathscr{B}(\tau+) = \mathscr{B}(0+)$, and $P_x\{X(\tau) = x\} = 1$. Let $A \in \mathscr{B}(0+)$ and $B \in \mathscr{B}$. Using strong Markov (1c),

$$\begin{aligned} P_x\{A \text{ and } B\} &= P_x\{A \text{ and } S_\tau \in B\} \\ &= \int_A P_{X(\tau)}\{B\}\, dP_x \\ &= \int_A P_x\{B\}\, dP_x \\ &= P_x\{A\} \cdot P_x\{B\}. \end{aligned}$$

Put $B = A$:

$$P_x\{A\} = P_x\{A\}^2 = 0 \text{ or } 1. \qquad \bigstar$$

Introduce the set

$$\begin{aligned} \{\text{stay at } x \text{ a bit}\} &= \{\omega : \omega \in \Omega \text{ and } \omega(s) = x \text{ for } 0 \leqq s \leqq \varepsilon \text{ with } \varepsilon = \varepsilon(\omega) > 0\}, \\ &= \textstyle\bigcup_{n=m}^{\infty} A_n, \end{aligned}$$

where

$$A_n = \{\omega : \omega \in \Omega \text{ and } \omega(r) = x \text{ for all rational } r \text{ with } 0 \leqq r \leqq 1/n\}.$$

Since $A_n \in \mathscr{B}(1/n)$, therefore $\{\text{stay at } x \text{ a bit}\} \in \mathscr{B}(0+)$. Use (4):

(5) Corollary. $P_x\{$*stay at x a bit*$\} = 0$ *or* 1.

Call x *instantaneous* iff $P_x\{\text{stay at } x \text{ a bit}\} = 0$ and *absorbing* iff $P_x\{\text{stay at } x \text{ a bit}\} = 1$.

(6) Fact. *If x is absorbing,* $P_x\{$*stay at x forever*$\} = 1$.

The set $\{\text{stay at } x \text{ forever}\}$ has only one element, namely the function which is identically x.

PROOF. Let $\tau(\omega)$ be the inf of rational $r > 0$ with $\omega(r) \neq x$, the inf of an empty set being ∞. Now

$$\{\tau < t\} = \{\omega : \omega \in \Omega \text{ and } \omega(r) \neq x \text{ for some rational } r < t\}$$

is in $\mathscr{B}(t)$, so τ is Markov. Check:

$$\{\tau > 0\} = \{\text{stay at } x \text{ a bit}\}$$

$$\{\tau = \infty\} = \{\text{stay at } x \text{ forever}\};$$

$$X(\tau) = x \quad \text{on } \{\tau < \infty\};$$

$$\tau(S_\tau) = 0 \quad \text{on } \{\tau < \infty\}.$$

Using strong Markov (1c),

$$\begin{aligned} P_x\{\tau < \infty\} &= P_x\{\tau < \infty \text{ and } \tau(S_\tau) = 0\} \\ &= \int_{\{\tau < \infty\}} P_{X(\tau)}\{\tau = 0\}\, dP_x \\ &= P_x\{\tau < \infty\} \cdot P_x\{\tau = 0\} \\ &= 0. \end{aligned}$$

★

Introduce

$$\begin{aligned} &\{\text{greater than } x \text{ near } 0\} \\ &\quad = \{\omega : \omega \in \Omega \text{ and } \omega(t) > x \text{ for } t\text{'s arbitrarily close to } 0\} \\ &\quad = \textstyle\bigcap_{n=m}^{\infty} A_n, \end{aligned}$$

where

$$A_n = \{\omega : \omega \in \Omega \text{ and } \omega(r) > x \text{ for some rational } r \text{ with } 0 \leqq r \leqq 1/n\}.$$

Since $A_n \in \mathscr{B}(1/n)$, therefore $\{\text{greater than } x \text{ near } 0\} \in \mathscr{B}(0+)$. Use (4):

(7) Corollary. $P_x\{$*greater than x near* $0\} = 0$ *or* 1.

Call x an *upper boundary* iff x is instantaneous and

$$P_x\{\text{greater than } x \text{ near } 0\} = 0.$$

(8) Corollary. *$P_x\{$less than x near $0\} = 0$ or 1.*

Call x a *lower boundary* iff x is instantaneous and $P_x\{\text{less than } x \text{ near } 0\} = 0$. Say x is *regular* iff

$$P_x\{\text{less than } x \text{ near } 0\} = P_x\{\text{greater than } x \text{ near } 0\} = 1.$$

The resulting classification is summarized in Figure 1.

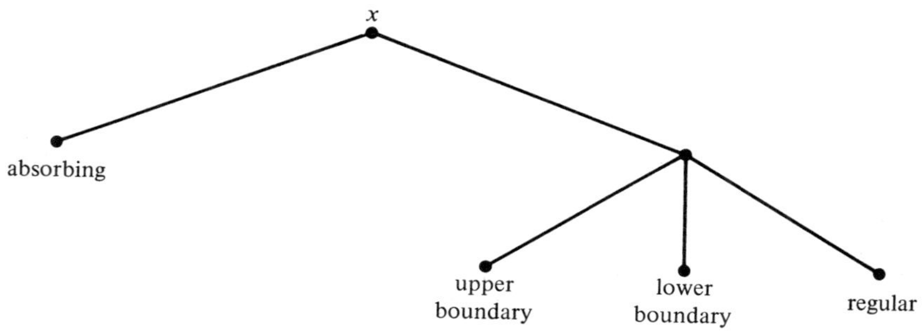

Figure 1.

(9) Definition. *$\{P_x\}$ is* regular *iff $P_x\{\text{hit } y\} > 0$ for all $x \in \text{int } I$ and all $y \in I$.*

As usual, int I is the interior of I. The connection between the classification of points and the classification of $\{P_x\}$ is established in (12), to which (10–11) are preliminary.

NOTATION. Let τ_x be the least t if any with $X(t) = x$, and $\tau_x = \infty$ if none. Thus, τ_x is Markov, and $\{\text{hit } x\} = \{\tau_x < \infty\}$ is in $\mathscr{B}(\tau_x +)$. Moreover,

$$X(\tau_x) = x \quad \text{on } \{\tau_x < \infty\}.$$

(10) Lemma. *$P_x\{\text{hit } z\} \geqq P_x\{\text{hit } y\} \cdot P_y\{\text{hit } z\}$, with equality provided $x \leqq y \leqq z$ or $x \geqq y \geqq z$.*

PROOF. Check

$$\{\tau_z < \infty\} \supset \{\tau_y < \infty\} \cap \{\tau_z(S_{\tau_y}) < \infty\}.$$

The first assertion now follows from the strong Markov property (1c). If y is between x and z, the continuity of ω makes

$$\{X(0) = x \text{ and } \tau_z < \infty\} = \{X(0) = x \text{ and } \tau_y < \infty\} \cap \{\tau_z(S_{\tau_y}) < \infty\}.$$

Use (1c) again. ★

For $a \neq b$, let

$$\{\text{hit } a \text{ before } b\} = \{\tau_a < \tau_b \leqq \infty\}.$$

I say $\{\text{hit } a \text{ before } b\} \in \mathscr{F}(\tau_a+)$. More generally,

(11) Lemma. *Let σ and τ be Markov times. Then*

(a) $\{\sigma < \tau\} \in \mathscr{B}(\sigma+)$

(b) $\{\sigma \leqq \tau\} \in \mathscr{B}(\sigma+)$

(c) $\{\sigma \leqq \tau\} \cap \mathscr{B}(\sigma+) \subset \mathscr{B}(\tau+)$.

(d) *if $\sigma \leqq \tau$ everywhere, then $\mathscr{B}(\sigma+) \subset \mathscr{B}(\tau+)$.*

PROOF. *Claim (a).* Check that

$$\{\sigma < \tau\} \cap \{\sigma < t\} = \bigcup_r \{\sigma < r < \tau\},$$

where r runs through the rationals in $[0, t)$. Plainly, $\{\sigma < r\} \in \mathscr{F}(r) \subset \mathscr{F}(t)$. Moreover,

$$\{\tau > r\} = \Omega \setminus \{\tau \leqq r\},$$

and for large m,

$$\{\tau \leqq r\} = \bigcap_{n=m}^{\infty} \left\{\tau < r + \frac{1}{n}\right\} \in \mathscr{B}(t).$$

Claim (b). Check that $\tau + \frac{1}{n}$ is again Markov, and

$$\{\sigma \leqq \tau\} = \bigcap_{n=1}^{\infty} \left\{\sigma < \tau + \frac{1}{n}\right\}.$$

Claim (c). Let $A \in \mathscr{B}(\sigma+)$ and $A \subset \{\sigma \leqq \tau\}$. Then

$$A \cap \{\tau < t\} = A \cap \{\sigma < t\} \cap \{\tau < t\}.$$

But $A \cap \{\sigma < t\} \in \mathscr{B}(t)$ and $\{\tau < t\} \in \mathscr{B}(t)$.

Claim (d). Use (c). ★

(12) Lemma. (a) *The right (left) endpoint of I is either not in I, or is in I and absorbing, or is in I and an upper (lower) boundary.*

(b) *If $\{P_x\}$ is regular, then each $x \in \text{int } I$ is regular.*

(c) *If each $x \in \text{int } I$ is regular, and for each endpoint y of I in I there is an $x = x(y) \in \text{int } I$ with $P_x\{\text{hit } y\} > 0$, then $\{P_x\}$ is regular.*

For example: if $I = (0, 1]$, then 0 is an endpoint of I not in I, and 1 is an endpoint of I in I.

PROOF. *Claim (a)* is immediate from the definitions.

Claim (b). As in (6), if x is an upper boundary, then

$$P_x\{X(t) \leqq x \text{ for all } t\} = 1.$$

Similarly for lower boundaries.

Claim (*c*). It is enough in view of (10) to check that $P_x\{\text{hit } x'\} > 0$ for all x and x' in int I. Suppose $x' > x$, the other case being symmetric. If $x < y_1 < y_2$ and $P_x\{\text{hit } y_2\} > 0$, then $P_x\{\text{hit } y_1\} > 0$ by (10), so

$$\{y : y \geqq x \text{ and } P_x[\text{hit } y] > 0\}$$

is a nonempty interval. Let z be the right endpoint of this interval. The problem is to prevent $z \in \text{int } I$. Suppose it were, and suppose $z > x$; the case $z = x$ is similar and easier.

Let

(13) $\quad L(t) = \min\{X(s) : 0 \leqq s \leqq t\} \quad$ and $\quad U(t) = \max\{X(s) : 0 \leqq s \leqq t\}$.

Because z is regular,

(14) $\quad L(t) < z < U(t) \quad$ for all $t > 0$ with P_z-probability 1.

I claim

(15) there are u and v with $x < u < z < v$ and $P_u\{\text{hit } v\} > 0$, as in Figure 2.

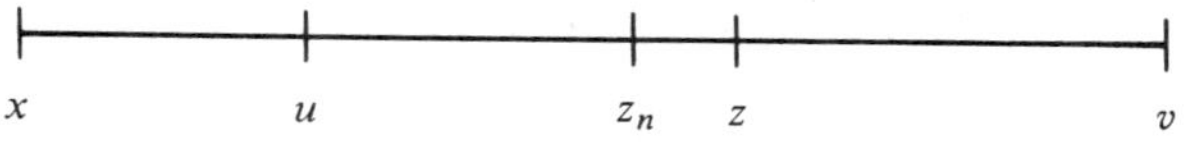

Figure 2.

Namely, use (14) to find $v > z$ with $P_z\{U(1) > v\} > 0$. Let $z_n = z - (1/n)$, and $\sigma_n = \tau_{z_n}$. Let A_n be the set $\{\sigma_n < \tau_v \leqq \infty\}$, and let B be the set $\{L(1) < z < v < U(1)\}$. Now $P_z\{B\} > 0$ by (14) and the choice of v. By path continuity,

$$\{X(0) = z \text{ and } B\} \setminus \{X(0) = z \text{ and } A_n\} \downarrow \phi,$$

so $P_z\{A_n\} > 0$ for large n. Check that

$$A_n = \{\sigma_n < \tau_v\} \cap \{S_{\sigma_n} \in [\text{hit } v]\}.$$

As (11) implies, $\{\sigma_n < \tau_v\} \in \mathscr{B}(\sigma_n +)$. Use strong Markov (1c) on σ_n:

$$P_z\{A_n\} = P_z\{\sigma_n < \tau_v \leqq \infty\} \cdot P_{z_n}\{\text{hit } v\}.$$

This proves (15) with $u = z_n$ for some large n. Then (10) and (15) imply $P_x\{\text{hit } v\} > 0$, a contradiction. ★

For the rest of this chapter, make the

(16) Assumption. $\{P_x\}$ IS A REGULAR DIFFUSION ON I.

Let $J = [a, b]$ be a closed, finite, proper interval included in I. This permits, for example, a and b to be endpoints of I in I.

NOTATION. Let $\tau_J = \tau_{ab}$ be the least t if any with $X(t) = a$ or b, and $\tau_J = \infty$ if none.

Clearly, τ_J is Markov.

(17) Lemma. *There is a $K < \infty$ and $\delta > 0$ such that $P_x\{\tau_J \leqq K\} \geqq \delta$ for all $x \in J$.*

PROOF. Let $c = \frac{1}{2}(a + b)$. By regularity, $P_c\{\tau_a < \infty\} > 0$ and $P_c\{\tau_b < \infty\} > 0$. Find $K < \infty$ so large that $P_c\{\tau_a \leqq K\} > 0$ and $P_c\{\tau_b \leqq K\} > 0$. Suppose $a \leqq x \leqq c$, as in Figure 3.

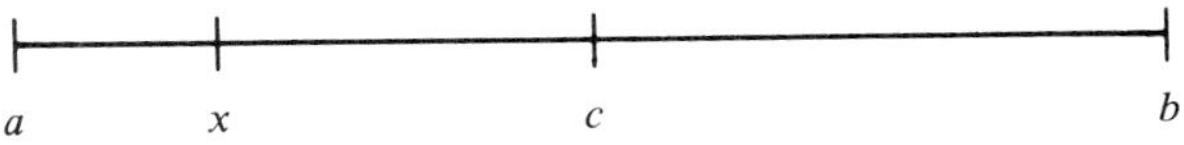

Figure 3.

By continuity,

$$\tau_a = \tau_x + \tau_a(S_{\tau_x}) \quad \text{on } \{X(0) = c\}.$$

So

$$\{X(0) = c \text{ and } \tau_a \leqq K\} \subset \{X(0) = c \text{ and } \tau_x < \infty \text{ and } \tau_a(S_{\tau_x}) \leqq K\}.$$

Using the strong Markov property (1c),

$$\begin{aligned} P_c\{\tau_a \leqq K\} &\leqq P_c\{\tau_x < \infty \text{ and } \tau_a(S_{\tau_x}) \leqq K\} \\ &= P_c\{\tau_x < \infty\} \cdot P_x\{\tau_a \leqq K\} \\ &\leqq P_x\{\tau_a \leqq K\}. \end{aligned}$$

Clearly, $\tau_{ab} \leqq \tau_a$, so $\{\tau_a \leqq K\} \subset \{\tau_{ab} \leqq K\}$, and

$$P_x\{\tau_{ab} \leqq K\} \geqq P_x\{\tau_a \leqq K\} \geqq P_c\{\tau_a \geqq K\}.$$

Similarly, if $c \leqq x \leqq b$,

$$P_x\{\tau_{ab} \leqq K\} \geqq P_c\{\tau_b \leqq K\}. \qquad \bigstar$$

(18) Corollary. *There are positive $r < 1$ and positive $A < \infty$ such that*

$$P_x\{\tau_J > t\} \leqq Ar^t$$

for all $x \in J$ and $t \in [0, \infty)$.

PROOF. Choose δ and K as in (17). Abbreviate

$$A_n = \{a < X(t) < b \text{ for } 0 \leqq t \leqq nK\}.$$

Check:

$$A_{n+1} = A_n \cap \{S_{nk} \in A_1\}; \quad \text{and} \quad A_n \in \mathscr{B}(nK+).$$

Moreover, $A_1 = \Omega \backslash \{\tau_{ab} \leqq K\}$; so $P_y\{A_1\} \leqq 1 - \delta$ for all $y \in [a, b]$. Use strong Markov (1c) on $\tau \equiv nK$ to check the next computation.

$$\begin{aligned} P_x\{\tau_J > (n+1)K\} &= P_x\{A_{n+1}\} \\ &= E_x\{1_{A_n} P_{X(nK)}(A_1)\} \\ &\leqq (1-\delta) P_x\{A_n\}. \end{aligned}$$

Induct:

$$P_x\{\tau_J > nK\} \leqq (1-\delta)^n.$$

Let $r = (1-\delta)^{1/K}$ and $A = 1/(1-\delta)$. Let $nK \leqq t \leqq (n+1)K$. Then

$$\begin{aligned} P_x\{\tau_J > t\} &\leqq P_x\{\tau_J > nK\} \\ &\leqq (1-\delta)^n \\ &= Ar^{(n+1)K} \\ &\leqq Ar^t. \end{aligned}$$

★

(19) Corollary. *$P_x\{\tau_J < \infty\} = 1$ and $E_x(\tau_J) < \infty$ for $x \in J$.*

The argument for (19) proves

(20) Lemma. *Suppose a is a lower boundary, $b \in I$, and $a \leqq x \leqq b$. Then $P_x\{\tau_b < \infty\} = 1$ and $E_x\{\tau_b\} < \infty$.*

By symmetry,

(21) Lemma. *Suppose b is an upper boundary, $a \in I$, and $a \leqq x \leqq b$. Then $P_x\{\tau_a < \infty\} = 1$ and $E_x\{\tau_a\} < \infty$.*

(22) Lemma. **(a)** *Suppose $x \in \operatorname{int} I$ and $b > x$. Then*

$$\lim_{a \uparrow x} P_x\{\text{hit } a \text{ before } b\} = 1.$$

(b) *Suppose $x \in \operatorname{int} I$ and $a < x$. Then*

$$\lim_{b \downarrow x} P_x\{\text{hit } b \text{ before } a\} = 1.$$

PROOF. *Claim (a).* Let $a_n \uparrow x$. Then $\{X(0) = x \text{ and hit } a_n \text{ before } b\}$ is nondecreasing, and the union of this sequence of sets includes, in the notation of (13),

$$\{X(0) = x \text{ and } L(t) < x < U(t) \text{ for all } t > 0\};$$

this set has P_x-probability 1 by (14).

Claim (b) is symmetric.

★

3. SCALE

Let $J = [a, b] \subset I$, with $a < b$. Let

$$s_J(x) = s_{ab}(x) = P_x\{\text{hit } b \text{ before } a\}.$$

As (18) implies, $P_x\{\text{hit } a \text{ before } b\} = 1 - s_J(x)$ for $x \in J$. Clearly, $s_J(a) = 0$ and $s_J(b) = 1$.

(23) Lemma. *s_J is strictly increasing.*

PROOF. Let $a \leqq x < y \leqq b$, as in Figure 4.

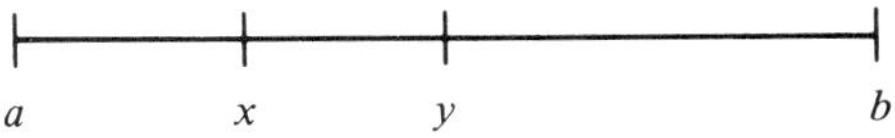

Figure 4.

I say

(24) $$P_x\{\text{hit } b \text{ before } a\} = P_x\{\text{hit } y \text{ before } a\} \cdot P_y\{\text{hit } b \text{ before } a\}.$$

Indeed, path continuity makes

$$\{X(0) = x \text{ and } \tau_b < \tau_a\} = \{X(0) = x \text{ and } \tau_y < \tau_a\} \cap S_{\tau_y}^{-1}\{\tau_b < \tau_a\}.$$

Use strong Markov (1c) on τ_y to get (24).

In particular, s_J is nondecreasing. I only need to get $s_J(x) < s_J(y)$ for $a < x < y < b$. Relation (24) reduces this to

(25) $$P_x\{\text{hit } y \text{ before } a\} < 1 \quad \text{for } a < x < y < b.$$

To begin with, $y \in \text{int } I$; so $P_y\{\text{hit } a\} > 0$ by assumption (16).

As in Figure 5, let σ_r be the least $t \geqq r$ if any with $X(t) = x$, and $\sigma_r = \infty$ if none. Let ρ_r be the least $t > \sigma_r$ if any with $X(t) = a$, and $\rho_r = \infty$ if none.

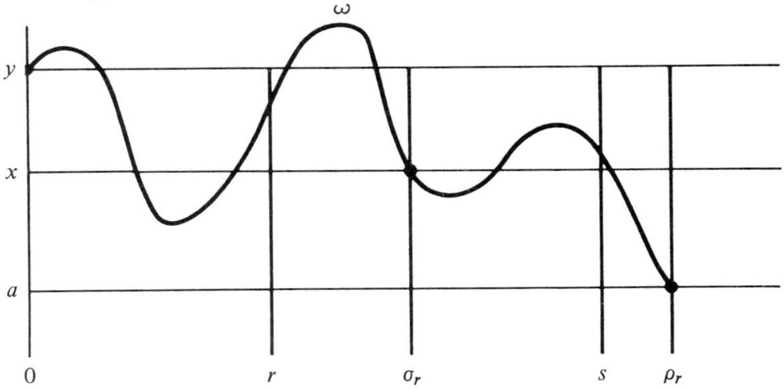

Figure 5.

You should check that σ_r and ρ_r are Markov. Let A_r be the set of $\omega \in \Omega$ with $\omega(0) = y$, and $\sigma_r(\omega) < \infty$, and $\rho_r(\omega) < \infty$, and $\omega(s) < y$ for $\sigma_r(\omega) \leqq s \leqq \rho_r(\omega)$.
By the continuity of paths,

$$\{X(0) = y \text{ and hit } a\} = \bigcup \{A_r : r \text{ is a nonnegative rational}\};$$

so $P_y\{A_r\} > 0$ for some nonnegative rational r, provided I get A_r to be measurable. But

$$A_r = \{X(0) = y \text{ and } \sigma_r < \infty\} \cap S_{\sigma_r}^{-1}\{\text{hit } a \text{ before } y\},$$

confirming measurability. The strong Markov property (1c) implies

$$P_y\{A_r\} = P_y\{\sigma_r < \infty\} \cdot P_x\{\text{hit } a \text{ before } y\}.$$

This forces $P_x\{\text{hit } a \text{ before } y\} > 0$, proving (25). ★

(26) Lemma. *s_J is continuous.*

PROOF. First, I will argue that s_J is continuous from the right at $x \in (a, b)$: the situation on the left is symmetric. Fix a sequence y_n in (x, b) which decreases to x. My job is to show that $s_J(y_n) \to s_J(x)$. As (24) implies,

$$s_J(x) = P_x\{\text{hit } y_n \text{ before } a\} \cdot s_J(y_n).$$

As (22b) implies, $P_x\{\text{hit } y_n \text{ before } a\} \to 1$, forcing $s_J(y_n) \to s_J(x)$.
Next, I will argue that s_J is continuous at b: the situation at a is symmetric. Fix $x \in (a, b)$ and $y_n \in (x, b)$ with $y_n \uparrow b$. My job is to show that $s_J(y_n) \to 1$. Let

$$B = \{\text{hit } b \text{ before } a\}$$

$$B_n = \{\text{hit } y_n \text{ before } a\}.$$

Check that $B_n \downarrow B$, on $\{X(0) = x \text{ and } \tau_{ab} < \infty\}$. Use (19):

$$P_x\{B_n\} \downarrow P_x\{B\}.$$

By (24),

$$P_x\{B\} = P_x\{B_n\} \cdot s_J(y_n).$$

So

$$s_J(y_n) = P_x\{B\}/P_x\{B_n\} \uparrow 1:$$

$P_x\{B_n\} > 0$ by (23). ★
Let $a \leqq a' \leqq x \leqq b' \leqq b$, as in Figure 6.

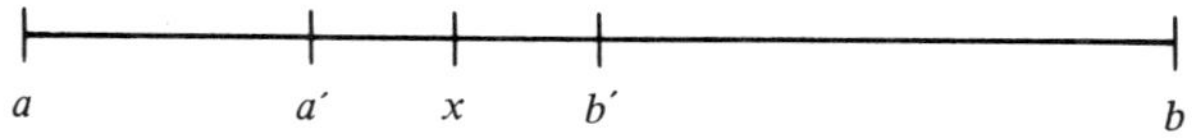

Figure 6.

I claim

(27) $$s_{ab}(x) = [1 - s_{a'b'}(x)] \cdot s_{ab}(a') + s_{a'b'}(x) \cdot s_{ab}(b').$$

Indeed, abbreviate $\sigma = \tau_{a'b'}$;

$$\begin{aligned} A &= \{X(0) = x \text{ and hit } a' \text{ before } b'\} \\ &= \{X(0) = x \text{ and } \sigma < \infty \text{ and } X(\sigma) = a'\}; \\ B &= \{X(0) = x \text{ and hit } b' \text{ before } a'\} \\ &= \{X(0) = x \text{ and } \sigma < \infty \text{ and } X(\sigma) = b'\}; \\ C &= \{\text{hit } b \text{ before } a\}. \end{aligned}$$

Then

$$\{X(0) = x \text{ and } C\} = \{A \text{ and } S_\sigma \in C\} \cup \{B \text{ and } S_\sigma \in C\}.$$

But A and B are in $\mathscr{B}(\sigma+)$. Use strong Markov (1c) and (19) to complete the proof of (27). ★

(28) Theorem. *There is a continuous, strictly increasing function S on I, such that for $a \leqq x \leqq b$ in I,*

$$s_{ab}(x) = P_x\{\textit{hit } b \textit{ before } a\} = \frac{S(x) - S(a)}{S(b) - S(a)}.$$

PROOF. Fix any proper, closed, finite interval $H = [\alpha, \beta] \subset I$. Let J, K be proper, closed, finite intervals with $H \subset J \subset K \subset I$. Define a function S_J on J as follows:

$$S_J = \frac{s_J - s_J(\alpha)}{s_J(\beta) - s_J(\alpha)}.$$

In particular, $S_H = s_H$. By (27), for some real numbers c and d,

$$s_K = cs_J + d \quad \text{on } J.$$

So for some real numbers c' and d',

$$S_K = c'S_J + d' \quad \text{on } J.$$

But $S_K(\alpha) = S_J(\alpha) = 0$ and $S_K(\beta) = S_J(\beta) = 1$, so $d' = 0$ and $c' = 1$. That is,

$$S_K = S_J \quad \text{on } J.$$

Consequently, there is a unique function S on I, whose retraction to $J \supset H$ is S_J. Relations (23) and (26) imply S is continuous and strictly increasing.

Finally, let $J = [a, b]$ be an arbitrary closed subinterval of I, and let K be a closed subinterval of I which includes both H and J. Now (27) implies that s_J is a linear function of s_K on J. By the first part of the argument, s_K is a linear function of S on K. Consequently, s_J is a linear function of

$$t_J = \frac{S - S(a)}{S(b) - S(a)}$$

on J. But $s_J = t_J$ at a and b, taking the values 0 and 1 there. Therefore, $s_J = t_J$ on J. ★

The function S is called the *scale function* of $\{P_x\}$. If $\hat{S}$ is another scale function, then $\hat{S}$ is a linear function of S. If $S(x) = x$ for all x, then $\{P_x\}$ is in *natural scale*.

Let Q_x be the $P_{S^{-1}x}$-distribution of the process $\{S[X(t)]:0 \leqq t < \infty\}$ for $x \in S(I)$.

(29) Theorem. *$\{Q_x\}$ is a diffusion on $S(I)$ in natural scale.*

PROOF. Use (3). ★

Recall that $BC(I)$ is the set of bounded, continuous, real-valued functions on I.

(30) Lemma. *For $f \in BC(I)$, the function $(x, t) \to E_x\{f[X(t)]\}$ is jointly continuous.*

PROOF. Fix $f \in BC(I)$, and $t \in [0, \infty)$, and $t_n \to t$, and $x \in \text{int } I$, and $x_n \to x$. The argument at an instantaneous boundary is similar. Abbreviate $\sigma_n = \tau_{x_n}$. By (14) and path continuity, $P_x\{\sigma_n \to 0\} = 1$. In particular, $P_x\{\sigma_n < \infty\} \to 1$. Dominated convergence makes

$$\int_{\{\sigma_n < \infty\}} f[X(\sigma_n + t_n)]\, dP_x \to E_x\{f[X(t)]\}. \tag{31}$$

By strong Markov (1d), the left side of (31) is

$$P_x\{\sigma_n < \infty\} \cdot E_{x_n}\{f[X(t_n)]\},$$

forcing

$$E_{x_n}\{f[X(t_n)]\} \to E_x\{f[X(t)]\}.$$

Let a be an absorbing left endpoint of I, the right endpoint case being symmetric. Let $f \in BC(I)$ with $f(a) = 0$, and let $x \to a$. I claim

$$E_x\{f[X(t)]\} \to 0 \quad \text{uniformly in } t \in [0, \infty). \tag{32}$$

Indeed, fix $\varepsilon > 0$. Choose $\delta_1 > 0$ so small that $a \leqq y \leqq a + \delta_1$ implies $|f(y)| \leqq \frac{1}{2}\varepsilon$. Remember a is absorbing. By (6) and strong Markov (1c),

$$P_x\{X(t) \neq a \text{ for some } t \geqq \tau_a\} = 0.$$

So

$$P_x\{\text{hit } (a + \delta_1) \text{ after } a\} = 0.$$

Use (26) to choose positive $\delta_2 < \delta_1$ so small that $a \leqq x \leqq a + \delta_2$ implies

$$P_x\{\text{hit } (a + \delta_1)\} = P_x\{\text{hit } (a + \delta_1) \text{ before } a\} \leqq \varepsilon/(2\|f\|).$$

Suppose $a \leqq x \leqq a + \delta_2$. Abbreviate $A = \{\text{hit } (a + \delta_1)\}$ and check this computation.

$$\begin{aligned} |E_x\{f[X(t)]\}| &\leqq \int_A |f[X(t)]|\, dP_x + \int_{\Omega \setminus A} |f[X(t)]|\, dP_x \\ &\leqq \|f\| P_x\{A\} + \max\{|f(y)| : a \leqq y \leqq a + \delta_1\} \\ &\leqq \varepsilon. \end{aligned}$$

★

Parenthetically, a Markov process with stationary transitions, continuous sample functions, and property (30) is strong Markov. This was the essence of (1.24).

4. SEMIGROUPS

Recall that $BC(I)$ is the space of bounded, continuous functions on I, and

$$\|f\| = \sup\{|f(x)| : x \in I\}.$$

For $0 \leqq t < \infty$, define the linear mapping T_t on $BC(I)$ by

(33) $$(T_t f)(x) = E_x\{f[X(t)]\}.$$

By (30),

(34) $$(t, x) \to (T_t f)(x) \quad \text{is continuous.}$$

Check

(35) $$\|T_t f\| \leqq \|f\|.$$

In particular, T_t maps $BC(I)$ into itself. I claim

(36) $$T_{t+s} = T_t T_s.$$

Indeed,

$$\begin{aligned} (T_{t+s} f)(x) &= E_x\{f[X(t + s)]\} \\ &= E_x\{E_{X(t)}\{f[X(s)]\}\} \\ &= [T_t(T_s f)](x), \end{aligned}$$

using strong Markov (1d) on $\tau \equiv t$ in line 2. Thus, $\{T_t\}$ is a one-parameter semigroup of contractions on $BC(I)$. Moreover, to restate (34–35),

(37) $t \to T_t f$ is continuous for uniform convergence on compact subsets of I, and is uniformly bounded on I.

Define the *Green operator* G_λ on $BC(I)$ for $\lambda > 0$:

$$(G_\lambda f)(x) = \int_0^\infty e^{-\lambda t}(T_t f \cdot x)\,dt. \tag{38}$$

Here and later, $g \cdot x = g(x)$. Typically, G_0 is meaningless. Clearly,

$$G_\lambda f \in BC(I) \quad \text{and} \quad \|G_\lambda f\| \leqq \|f\|/\lambda. \tag{39}$$

(40) Lemma. (*Resolvent equation*). $G_\lambda - G_\mu + (\lambda - \mu)G_\lambda G_\mu = 0$.

PROOF. Suppose $\lambda \neq \mu$ and $f \in BC(I)$. Then

$$\begin{aligned}
G_\lambda G_\mu f &= \int_0^\infty e^{-\lambda t}\, T_t G_\mu f\, dt \\
&= \int_0^\infty e^{-\lambda t}\, T_t\left(\int_0^\infty e^{-\mu s}\, T_s f\, ds\right) dt \\
&= \int_0^\infty \int_0^\infty e^{-\lambda t - \mu s}\, T_t T_s f\, ds\, dt && \text{by Fubini} \\
&= \int_0^\infty \int_0^\infty e^{-\lambda t - \mu s}\, T_{t+s} f\, ds\, dt && \text{by (36)} \\
& && \text{(put } v = t + s \text{ in the inner integral)} \\
&= \int_0^\infty \int_t^\infty e^{-(\lambda-\mu)t}\, e^{-\mu v}\, T_v f\, dv\, dt \\
&= \int_0^\infty \int_0^v e^{-(\lambda-\mu)t}\, e^{-\mu v}\, T_v f\, dt\, dv && \text{by Fubini} \\
&= \frac{1}{\mu - \lambda}\int_0^\infty (e^{-\lambda v} - e^{-\mu v}) T_v f\, dv \\
&= \frac{1}{\mu - \lambda}(G_\lambda - G_\mu) f.
\end{aligned}$$

★

(41) Corollary. *G_λ and G_μ commute.*

PROOF. Use (40). ★

(42) Corollary. *$G_\lambda f \equiv 0$ iff $G_\mu f \equiv 0$.*

PROOF. Use (40). ★

(43) Corollary. *$G_\lambda f \equiv 0$ iff $f \equiv 0$.*

PROOF. Suppose $G_\lambda f \equiv 0$. Use (42) and (37) to see that as $\mu \to \infty$,

$$0 = \mu G_\mu f = \int_0^\infty \mu e^{-\mu t} T_t f \, dt = \int_0^\infty e^{-s} T_{s/\mu} f \, ds \to f. \quad ★$$

(44) Corollary. *The range of G_λ does not depend on λ.*

PROOF. Suppose $f = G_\mu g$. Then (40) makes

$$f = G_\lambda g + (\lambda - \mu) G_\lambda G_\mu g = G_\lambda[g + (\lambda - \mu) G_\mu g]. \quad ★$$

(45) Definition. *Δ is the common range of G_λ. This is legitimate by* (44). *On Δ, let $\Gamma = 1 - G_1^{-1}$. This is legitimate by* (43) *and linearity. Then* (Γ, Δ) *is the* infinitesimal generator *of* $\{P_x\}$ *or* $\{T_t\}$.

(46) Lemma. *$G_\mu^{-1} = \mu - \Gamma$ on Δ.*

PROOF. Apply G_1^{-1} to (40) with $\lambda = 1$.

(47) Theorem. *(Γ, Δ) determines $\{P_x\}$.*

PROOF. As (46) implies, (Γ, Δ) determines G_λ for all λ. By the uniqueness theorem for Laplace transforms (3.70), this determines T_t for all t. I say this determines

$$E_x\{f_1[X(t_1)] f_2[X(t_2)] \cdots f_n[X(t_n)]\}$$

for all $t_1 < t_2 < \cdots < t_n$ and $f_1, \ldots, f_n \in BC(I)$. The case $n = 1$ is clear, from definition (33). Let $t_{n+1} = t_n + s$ and

$$Y_{n-1} = f_1[X(t_1)] \cdots f_{n-1}[X(t_{n-1})]$$

and

$$g(x) = f_n(x) E_x\{f_{n+1}[X(s)]\}.$$

Use strong Markov (1d) on $\tau \equiv t_n$:

$$E_x\{Y_{n-1} f_n[X(t_n)] f_{n+1}[X(t_n + s)]\} = E_x\{Y_{n-1} g[X(t_n)]\}.$$

This reduces the case $n + 1$ to the case n. Induct. ★

(48) Lemma. (*Dynkin's formula*). *If τ is Markov, $E_x\{\tau\} < \infty$, and $u \in \Delta$, then*

$$E_x\left\{\int_0^\tau \Gamma u \cdot X(t) \, dt\right\} = E_x\{u[X(\tau)]\} - u(x).$$

PROOF. Let $f \in BC(I)$ and $u = G_\alpha f$. For any Markov τ, I claim

$$u(x) = E_x\left\{\int_0^\tau e^{-\alpha t} f[X(t)]\,dt\right\} + E_x\{e^{-\alpha\tau} u[X(\tau)]\}. \tag{49}$$

Indeed,

$$\begin{aligned}
u(x) &= G_\alpha f \cdot x \\
&= \int_0^\infty e^{-\alpha t}\, T_t f \cdot x\,dt \\
&= \int_0^\infty e^{-\alpha t}\, E_x\{f[X(t)]\}\,dt \\
&= E_x\left\{\int_0^\infty e^{-\alpha t} f[X(t)]\,dt\right\} \\
&= E_x\left\{\int_0^\tau e^{-\alpha t} f[X(t)]\,dt\right\} + E_x\left\{\int_\tau^\infty e^{-\alpha t} f[X(t)]\,dt\right\}.
\end{aligned}$$

To evaluate the second term, put $s = t - \tau$:

$$\begin{aligned}
E_x\left\{\int_\tau^\infty e^{-\alpha t} f[X(t)]\,dt\right\} &= E_x\left\{e^{-\alpha\tau}\int_0^\infty e^{-\alpha s} f[X(\tau + s)]\,ds\right\} \\
&= E_x\{e^{-\alpha\tau} u[X(\tau)]\},
\end{aligned}$$

using strong Markov (1d).

Fix $u \in \Delta$. Let $f_\alpha = (\alpha - \Gamma)u \in BC(I)$, so $u = G_\alpha f_\alpha$ by (46). Put f_α for f in (49):

$$\begin{aligned}
u(x) &= E_x\left\{\int_0^\tau e^{-\alpha t}[(\alpha - \Gamma)u] \cdot X(t)\,dt\right\} + E_x\{e^{-\alpha\tau} u[X(\tau)]\} \\
&= E_x\left\{\int_0^\tau \alpha e^{-\alpha t} u \cdot X(t)\,dt\right\} - E_x\left\{\int_0^\tau e^{-\alpha t}\Gamma u \cdot X(t)\,dt\right\} \\
&\qquad + E_x\{e^{-\alpha\tau} u[X(\tau)]\}.
\end{aligned} \tag{50}$$

Let $\alpha \downarrow 0$. The third term converges to $E_x\{u[X(\tau)]\}$ by dominated convergence. Because $E_x\{\tau\} < \infty$, dominated convergence works on the first two terms. The first one goes to 0, and the second to $E_x\{\int_0^\tau \Gamma u \cdot X(t)\,dt\}$. ★

Suppose x is instantaneous. Define N and σ_N as follows. For $x \in \text{int}\, I$, let N be a closed interval containing x as an interior point; and let $\sigma_N = \tau_N$ be the time to hit an endpoint of N. If x is a lower boundary, let $N = [x, x + \varepsilon]$ for some $\varepsilon > 0$; and let $\sigma_N = \tau_{x+\varepsilon}$ be the time to hit $x + \varepsilon$. If x is an upper boundary, let $N = [x - \varepsilon, x]$; and let $\sigma_N = \tau_{x-\varepsilon}$ be the time to hit $x - \varepsilon$.

(51) Corollary. *If $u \in \Delta$ and x is instantaneous,*

$$\Gamma u \cdot x = \lim_{N \downarrow x} \frac{E_x\{u[X(\sigma_N)]\} - u(x)}{E_x\{\sigma_N\}}.$$

PROOF. Use one of (19, 20, 21) to check $E_x\{\sigma_N\} < \infty$. Then use (48) with $\tau = \sigma_N$. Divide both sides of the equation by $E_x(\sigma_N)$. If $X(0) = x$ and $0 \leqq t \leqq \sigma_N$, then $X(t) \in N$; so $\Gamma u \cdot X(t)$ is nearly $\Gamma u \cdot x$ for small N. More formally, fix $\delta > 0$. Remember $\Gamma u \in BC(I)$. Choose N so small that

$$(\Gamma u \cdot x) - \delta < \Gamma u \cdot y < (\Gamma u \cdot x) + \delta$$

for all $y \in N$. Then

$$[(\Gamma u \cdot x) - \delta]E_x\{\sigma_N\} \leqq E_x\left\{\int_0^{\sigma_N} \Gamma u \cdot X(t)\, dt\right\} \leqq [(\Gamma u \cdot x) + \delta]E_x(\sigma_N). \quad ★$$

(52) Fact. *If $u \in \Delta$ and x is absorbing, $\Gamma u \cdot x = 0$.*

PROOF. Let $u = G_1 f$ for $f \in BC(I)$, so $\Gamma u = G_1 f - f$ by (45). But $T_t f \cdot x = f(x)$ for all t, so $G_1 f \cdot x = f(x)$. ★

Let $\Gamma_\varepsilon = \varepsilon^{-1}(T_\varepsilon - 1)$ for $\varepsilon > 0$. Let $\Gamma_0 = \lim_{\varepsilon \to 0} \Gamma_\varepsilon$ on the set Δ_0 of $f \in BC(I)$ where $\Gamma_\varepsilon f$ converges boundedly and pointwise to an element of $BC(I)$. Let $\Gamma_{00} = \lim_{\varepsilon \to 0} \Gamma_\varepsilon$ on the set Δ_{00} of $f \in BC(I)$ where $\Gamma_\varepsilon f$ converges boundedly and uniformly on compact sets to an element of $BC(I)$.

(53) Theorem. $\Delta = \Delta_0 = \Delta_{00}$ *and* $\Gamma = \Gamma_0 = \Gamma_{00}$.

PROOF. Let $u \in \Delta$. Use Dynkin's formula (48) with $\tau \equiv \varepsilon$:

$$(\Gamma_\varepsilon u)(x) = \varepsilon^{-1} \int_0^\varepsilon E_x[\Gamma u \cdot X(t)]\, dt \to \Gamma u \cdot x,$$

boundedly and uniformly on compacts by (37). So, $\Delta \subset \Delta_{00}$, and $\Gamma = \Gamma_{00}$ on Δ. Of course, $\Delta_{00} \subset \Delta_0$, and $\Gamma_0 = \Gamma_{00}$ on Δ_{00}. It remains only to see $\Delta_0 \subset \Delta$. So, let $u \in \Delta_0$. Then $\Gamma_0 u \in BC(I)$ and

$$\begin{aligned}
G_1(1 - \Gamma_0)u &= G_1 u - \lim_{\varepsilon \to 0} G_1 \Gamma_\varepsilon u && \text{by dominated convergence}\\
&= G_1 u - \lim_{\varepsilon \to 0} \varepsilon^{-1} G_1(T_\varepsilon u - u)\\
&= G_1 u - \lim_{\varepsilon \to 0} \varepsilon^{-1} \int_0^\infty e^{-t}(T_{t+\varepsilon} u - T_t u)\, dt && \text{by (36)}\\
&= G_1 u - \lim_{\varepsilon \to 0} \varepsilon^{-1}\left[(e^\varepsilon - 1)\int_\varepsilon^\infty e^{-t}\, T_t u\, dt - \int_0^\varepsilon e^{-t}\, T_t u\, dt\right]\\
&= G_1 u - G_1 u + u && \text{by calculus and (37)}\\
&= u.
\end{aligned}$$

So $u = G_1(1 - \Gamma_0)u \in \Delta$. ★

The next lemma will be useful for identifying Γ in Section 7.

(54) Lemma *Let Δ_0 be a linear subset of $BC(I)$, and Γ_0 a linear mapping of Δ_0 to $BC(I)$. Suppose (Γ_0, Δ_0) is an extension of (Γ, Δ), and $\Gamma_0 v = v$ implies $v = 0$. Then $\Delta_0 = \Delta$.*

NOTE. $\Gamma v = v$ implies $G_1^{-1} v = (1 - \Gamma)v = 0$, so $v = G_1 0 = 0$. And the (Γ_0, Δ_0) of (54) has no a priori connection with the (Γ_0, Δ_0) of (53).

PROOF. Suppose $u \in \Delta_0$. Then $(1 - \Gamma_0)u \in BC(I)$, and $G_1(1 - \Gamma_0)u \in \Delta$, so $\Gamma_0 = \Gamma$ at this function. Let

$$v = u - G_1(1 - \Gamma_0)u \in \Delta_0.$$

Now

$$\begin{aligned} \Gamma_0 v &= \Gamma_0 u - \Gamma_0 G_1(1 - \Gamma_0)u \\ &= \Gamma_0 u - \Gamma G_1(1 - \Gamma_0)u \\ &= \Gamma_0 u - (1 - G_1^{-1})G_1(1 - \Gamma_0)u \\ &= \Gamma_0 u - G_1(1 - \Gamma_0)u + (1 - \Gamma_0)u \\ &= v. \end{aligned}$$

So $v = 0$. That is, $u = G_1(1 - \Gamma_0)u \in \Delta$. ★

5. GREEN'S FUNCTION

Let $a < b$ be finite real numbers, and let μ be a locally finite measure on (a, b); that is, μ assigns finite mass to compact subsets of (a, b). As usual, g^+ is the right derivative of g, and g^- the left derivative. Moreover, *g is BV(loc) on (a, b)* means that g is of bounded variation on $[a + \varepsilon, b - \varepsilon]$ for any ε with $0 < \varepsilon < \frac{1}{2}(b - a)$. Let α and β be real numbers, and consider the following conditions on f, a real-valued function on $[a, b] = J$.

(55.1) $f(a) = \alpha, \quad f(b) = \beta, \quad$ and $\quad f$ is continuous.

(55.2) f^+ exists, is finite, continuous from the right, and BV(loc), on (a, b).

(55.3) $df^+ = d\mu \quad$ on (a, b).

Condition (55.2) is silent on the existence of f^+ at a. Condition (55.3) means that for $a < x \leqq y < b$,

$$f^+(y) - f^+(x) = \mu(x, y].$$

Measures are nonnegative: so f^+ is nondecreasing, and f is convex Consider the following condition on μ.

(56) For ε with $0 < \varepsilon < b - a$,

$$\int_{b-\varepsilon}^{b} (b - y)\,\mu(dy) < \infty \quad \text{and} \quad \int_{a}^{a+\varepsilon} (y - a)\,\mu(dy) < \infty.$$

(57) Definition. *The* Green's function $G_{ab} = G_J$ *on* $[a, b] \times [a, b]$ *is*

$$G_{ab}(x, y) = \frac{(x - a)(b - y)}{b - a} \quad \text{for } a \leqq x \leqq y \leqq b$$

$$= \frac{(y - a)(b - x)}{b - a} \quad \text{for } a \leqq y \leqq x \leqq b.$$

This is illustrated in Figure 7.

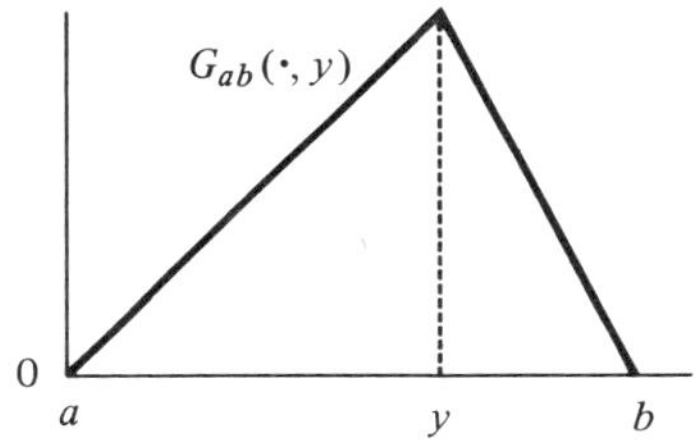

Figure 7.

(58) Lemma. (**a**) *If* (56) *holds, then*

(59) $$f(x) = \alpha\frac{b - x}{b - a} + \beta\frac{x - a}{b - a} - \int_J G_J(x, y)\mu(dy)$$

satisfies (55).

(**b**) *If f satisfies* (55), *then μ satisfies* (56) *and f is given by* (59).

PROOF. *Claim* (*a*). Without real loss, suppose $a = 0$ and $b = 1$. Abbreviate G for G_{01}. Suppose (56). I will check that f defined by (59) satisfies (55). To begin with, $f(0) = \alpha$ and $f(1) = \beta$, because $G(0, \cdot) = G(1, \cdot) = 0$. Next, $G(\cdot, y) \leqq G(y, y)$, which is μ-summable by (56); and $G(\cdot, y)$ is continuous: so dominated convergence makes f continuous, and (55.1) holds. Let

$$\varphi(x) = \int_{[0,1]} G(x, y)\,\mu(dy).$$

I have to show that $-\varphi$ satisfies (55.2) and (55.3).

Let $0 < x < 1$. Let $\varepsilon > 0$ and $\varepsilon \to 0$, with $x + \varepsilon < 1$. Let

$$A = \int_{(0,x]} y\,\mu(dy) \quad \text{and} \quad B = \int_{(x,1]} (1 - y)\,\mu(dy).$$

Then

$$\varphi(x) = (1 - x)A + xB.$$

Let

$$A^* = \int_{(x,x+\varepsilon]} y\,\mu(dy) \quad \text{and} \quad B^* = \int_{(x,x+\varepsilon]} (1 - y)\,\mu(dy).$$

Then

$$\begin{aligned}\varphi(x + \varepsilon) &= (1 - x - \varepsilon)(A + A^*) + (x + \varepsilon)(B - B^*)\\ &= \varphi(x) - \varepsilon A + \varepsilon B + T\end{aligned}$$

where

$$\begin{aligned}T &= (1 - x - \varepsilon)A^* - (x + \varepsilon)B^*\\ &= [A^* - x(A^* + B^*)] - \varepsilon(A^* + B^*)\\ &= \int_{(x,x+\varepsilon]} (y - x)\,\mu(dy) - \varepsilon \int_{(x,x+\varepsilon]} \mu(dy)\\ &= o(\varepsilon).\end{aligned}$$

So,

$$\varphi^+(x) = -\int_{(0,x]} y\,\mu(dy) + \int_{(x,1]} (1 - y)\,\mu(dy)$$

and

$$d\varphi^+ = -d\mu.$$

Claim (*b*). Suppose f satisfies (55). My problem is to prove (56) and (59). A glance at Figure 8 may help.

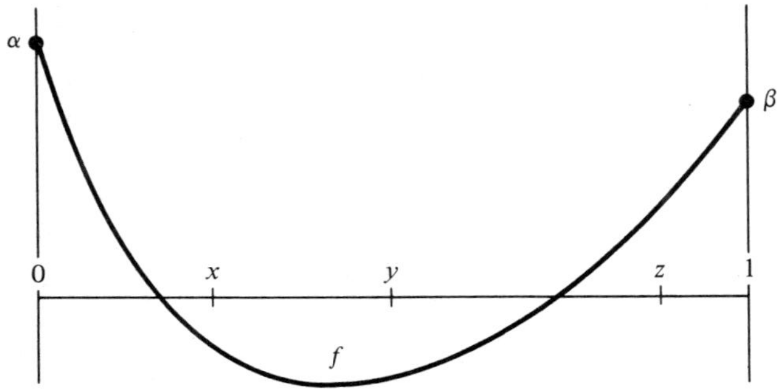

Figure 8.

Let $0 < x \leqq z \leqq y < 1$. Then

$$f(y) - f(x) = \int_x^y f^+(z)\,dz$$

by (55.2) and Dini (3.69). Let $y \uparrow 1$. Use (55.1) to evaluate the left hand limit. Remember that f^+ is monotone; use monotone convergence to evaluate the right hand limit. The upshot is

$$f(1) - f(x) = \int_x^1 f^+(z)\,dz.$$

Use (55.3) and Fubini:

$$\begin{aligned} f(1) - f(x) &= \int_x^1 \{f^+(x) + \int_{(x,z]} \mu(dw)\}\,dz \\ &= (1 - x)f^+(x) + \int_x^1 \int_w^1 dz\,\mu(dw). \end{aligned}$$

That is,

$$(\alpha) \qquad f(1) - f(x) = (1 - x)f^+(x) + \int_x^1 (1 - w)\,\mu(dw).$$

In particular, the last integral is finite, proving half of (56). Similarly,

$$(\beta) \qquad f(x) - f(0) = xf^+(x) - \int_0^x w\,\mu(dw),$$

proving the other half of (56). Multiply (α) by $-x$ and (β) by $(1 - x)$; add to get

$$f(x) = xf(1) + (1 - x)f(0) - \int_{(0,1)} G(x, y)\,\mu(dy). \qquad \bigstar$$

(60) Corollary. *Suppose* (56), *and define* f *by* (59).

(a) f^+ *exists and is continuous even at* a, *but* $f^+(a) = -\infty$ *is possible. Indeed,* $f^+(a) = -\infty$ *iff* $\mu(a, a + \varepsilon) = \infty$ *for* ε *with* $0 < \varepsilon < b - a$.

(b) f^- *exists, is finite, continuous from the left, and* $BV(\text{loc})$, *on* (a, b). *Moreover,* f^- *exists and is continuous, even at* b, *but* $f^-(b) = \infty$ *is possible. Indeed,* $f^-(b) = \infty$ *iff* $\mu(b - \varepsilon, b) = \infty$ *for* ε *with* $0 < \varepsilon < b - a$. *Furthermore,* $df^- = d\mu$ *on* (a, b). *And* $f^+ = f^-$ *on* (a, b) *except at* x *with* $\mu\{x\} > 0$.

PROOF. *Claim* (a). Let $0 < x < y < 1$. By (55.3),

$$f^+(y) - f^+(x) = \mu(x, y].$$

Let $x \downarrow a$ and use (3.74h):

$$f^+(u) - f^+(a) = \mu(a, y].$$

Claim (b). Use (3.74c, e). ★

6. SPEED MEASURE

Recall that E_x is expectation with respect to P_x. From now on, make the

(61) Assumption. $\{P_x\}$ IS A REGULAR DIFFUSION IN NATURAL SCALE ON I.

Let $J = [a, b] \subset I$. Recall that $\tau_J = \tau_{ab}$ is the least t, if any, such that $X(t) = a$ or b, and $\tau_J = \infty$ if none. Let

$$e_{ab}(x) = e_J(x) = E_x\{\tau_J\}$$

for $x \in J$, finite by (19). Of course, $e_J(a) = e_J(b) = 0$.

(62) Theorem. *e_J is continuous and strictly concave on J, as in Figure 9.*

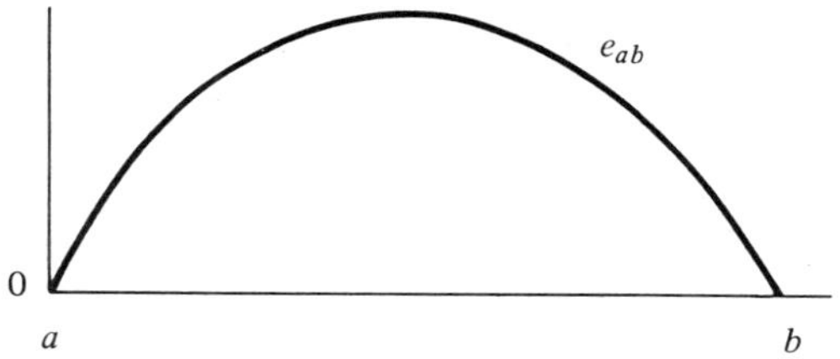

Figure 9.

PROOF. Let $a \leqq \alpha \leqq x \leqq \beta \leqq b$, as in Figure 10, and let $K = [\alpha, \beta]$.

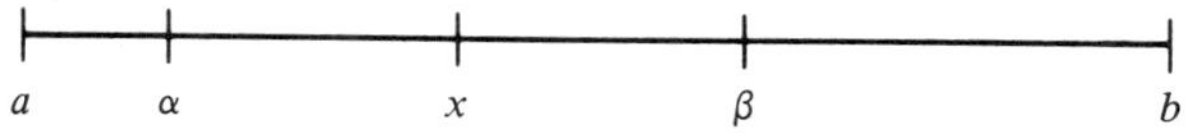

Figure 10.

I claim

$$(63) \qquad E_x\{\tau_J\} = E_x\{\tau_K\} + \frac{x - \alpha}{\beta - \alpha} \cdot E_\beta\{\tau_J\} + \frac{\beta - x}{\beta - \alpha} \cdot E_\alpha\{\tau_J\}.$$

Indeed, put $\sigma = \tau_K$;

$$A = \{X(0) = x \text{ and hit } \alpha \text{ before } \beta\}$$
$$= \{X(0) = x \text{ and } \sigma < \infty \text{ and } X(\sigma) = \alpha\};$$
$$B = \{X(0) = x \text{ and hit } \beta \text{ before } \alpha\}$$
$$= \{X(0) = x \text{ and } \sigma < \infty \text{ and } X(\sigma) = \beta\}.$$

By (61),

$$P_x\{A\} = \frac{\beta - x}{\beta - \alpha} \quad \text{and} \quad P_x\{B\} = \frac{x - \alpha}{\beta - \alpha}.$$

Check that

$$\tau_J = \sigma + \tau_J(S_\sigma) \quad \text{on } \{X(0) = x\}.$$

But A and B are in $\mathscr{B}(\sigma+)$, and $A \cup B$ is P_x-almost all of Ω. So

$$E_x\{\tau_J\} = E_x\{\sigma\} + \int_A \tau_J(S_\sigma)\,dP_x + \int_B \tau_J(S_\sigma)\,dP_x.$$

Now strong Markov (1d) does (63). Plainly, $E_x\{\tau_K\} > 0$ for $\alpha < x < \beta$. So e_J is strictly concave on (a, b). In particular, e_J is continuous on (a, b) by (3.74a).

To show e_J is continuous at a, fix $x \in (a, b)$ and let $a_n \downarrow a$ with $a_1 < x$, as in Figure 11.

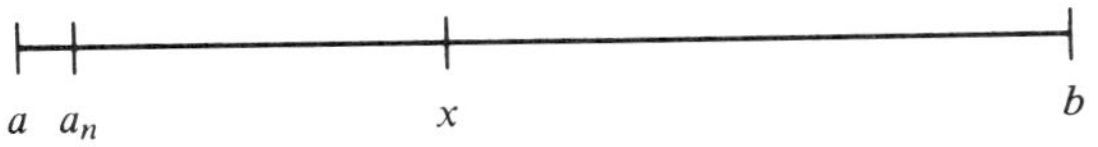

Figure 11.

Then $\tau_{a_n b} \uparrow \tau_{ab}$ on $\{X(0) = x$ and $\tau_{ab} < \infty\}$. So $E_x\{\tau_{a_n b}\} \uparrow E_x\{\tau_{ab}\}$. As a special case of (63),

$$E_x\{\tau_{ab}\} = E_x\{\tau_{a_n b}\} + \frac{b - x}{b - a_n} \cdot E_{a_n}\{\tau_{ab}\}.$$

Therefore, $E_{a_n}\{\tau_{ab}\} \to 0$. The argument for b is symmetric and is omitted. ★

(64) Corollary. **(a)** *e_J^+ exists, is continuous from the right, and is strictly decreasing, on $[a, b)$.*

(b) *e_J^- exists, is continuous from the left, and is strictly decreasing, on $(a, b]$.*

(c) *On (a, b): the discontinuity set D of e_J^+ and e_J^- is the same, and is countable. And $e_J^+ = e_J^-$ on $(a, b) \setminus D$. On D,*

$$e_J^+(x-) - e_J^+(x) = e_J^-(x) - e_J^-(x+).$$

(d) *Moreover, $de_J^+ = de_J^-$ on (a, b).*

PROOF. Use (62) and (3.74). ★

(65) Corollary. *Let K be a closed subinterval of J. Then $e_J^+ = e_K^+ + $ constant and $de_J^+ = de_K^+$ on* int K.

PROOF. Use (63). ★

(66) Theorem. *There is a unique measure m on* int I, *whose retract to* int J *has distribution function $-e_J^+$, for all intervals $J = [a, b] \subset I$. This m is called the* speed measure *of $\{P_x\}$. Speed measure is locally finite, that is, it assigns finite mass to compact subsets of* int I. *Speed measure is strictly positive, that is, it assigns positive mass to all nonempty open subsets of* int I.

PROOF. Use (64–65). ★

NOTATION. UNTIL FURTHER NOTICE, m IS THE SPEED MEASURE OF $\{P_x\}$.

Recall the Green's function G_J from (57).

(67) Corollary. **(a)** $\int_J (x - a)(b - x)\, m(dx) < \infty$.

(b) $e_J(x) = \int_J G_J(x, y)\, m(dy)$ *for $x \in J$.*

PROOF. Use (58) with $f = -e_J$. ★

(68) Lemma. *Let $J = [a, b] \subset I$. Let $x \in J$. Let f be a bounded, measurable function on I. Then*

$$E_x\left\{\int_0^{\tau_J} f[X(t)]\, dt\right\} = \int_J G_J(x, y) f(y)\, m(dy).$$

PROOF. It is enough to do $J = [0, 1]$. Write τ for τ_J, and G for G_J. Both sides of the equation in (68) are linear in f, and continuous under uniformly bounded pointwise passages to the limit: $E_x\{\tau\} < \infty$ by (19), handling the continuity of the left side. Both sides of the equation depend only on f retracted to J. Finally, (68) holds for $f \equiv 1$ by (67b). By a standard approximation argument, it is enough to prove (68) for $f = 1_{[c,1]}$ with $0 < c < 1$.

Suppose $0 \leqq g \leqq 1$ is measurable. Let

$$\hat{g}(x) = E_x\left\{\int_0^{\tau} g[X(t)]\, dt\right\}.$$

Let $0 \leqq \alpha \leqq x \leqq \beta \leqq 1$, as in Figure 12.

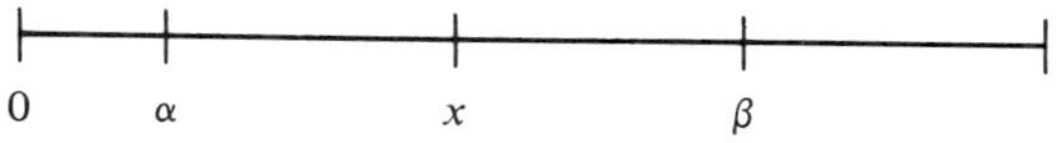

Figure 12.

As for (63),

$$\hat{g}(x) = h(x) + \frac{x-\alpha}{\beta-\alpha}\hat{g}(\beta) + \frac{\beta-x}{\beta-\alpha}\hat{g}(\alpha), \tag{69}$$

where

$$h(x) = E_x\left\{\int_0^{\tau_{\alpha\beta}} g[X(t)]\,dt\right\} \geqq 0.$$

So $\hat{g}$ is concave on J. And $\hat{g}$ vanishes continuously at 0 and 1, because $\hat{g} \leqq e_J$. Suppose $0 \leqq \alpha \leqq x \leqq 1$, as in Figure 12. Suppose $g \equiv 1$ on $(\alpha, 1]$. As a special case of (69),

$$\hat{g}(x) = E_x\{\tau_{\alpha 1}\} + \frac{1-x}{1-\alpha}\hat{g}(\alpha).$$

Consequently, $d\hat{g}^+ = -dm$ on $(\alpha, 1)$. Now use (58). Let $f = -\hat{g}$ and $d\mu = -d\hat{g}^+$. Then f satisfies (55) with $\alpha = \beta = 0$. As (58b) implies, μ satisfies (56). By (58a),

$$\begin{aligned}
-\hat{g} &= -\int_{(0,1]} G(\cdot, y)\,\mu(dy)\\
&\leqq -\int_{(\alpha,1]} G(\cdot, y)\,\mu(dy)\\
&= -\int_{(\alpha,1]} G(\cdot, y)\,m(dy).
\end{aligned}$$

Put $g = 1_{(\alpha,1]}$ and multiply by -1:

$$\hat{1}_{(\alpha,1]} \geqq \int_{(\alpha,1]} G(\cdot, y)\,m(dy).$$

Let α increase to c:

$$\hat{1}_{[c,1]} \geqq \int_{[c,1]} G(\cdot, y)\,m(dy). \tag{70a}$$

Similarly,

$$\hat{1}_{[0,c)} \geqq \int_{[0,c)} G(\cdot, y)\,m(dy). \tag{70b}$$

Add (70a–b); use (67b) to see that equality holds in (70a–b). ★

(71) Corollary. *If $J = [a, b] \subset I$ and $x \in J$ and $f \in \Delta$, then*

$$\int_J G_J(x, y)\Gamma f \cdot (y)\,m(dy) = \frac{x-a}{b-a}f(b) + \frac{b-x}{b-a}f(a) - f(x).$$

PROOF. Use (68) with Γf for f; and Dynkin's formula (48) with $\tau = \tau_{ab}$ and $u = f$. This choice of τ is legitimate by (19). ★

(72) Lemma. *If $J = [a, b] \subset I$, and $f \in \Delta$, then f^+ exists, is finite, continuous from the right, and BV(loc), on (a, b). Moreover, $df^+ = \Gamma f\, dm$ on (a, b).*

PROOF. As definition (45) implies, $\Gamma f \in BC(I)$. Let $g = \max\{\Gamma f, 0\}$ and $h = g - \Gamma f \geqq 0$. So g and h are the positive and negative parts of Γf, and are in $BC(I)$. Clearly, $\Gamma f = g - h$. Let

$$\lambda(x) = \frac{x - a}{b - a} f(b) + \frac{b - x}{b - a} f(a)$$

$$\gamma(x) = -\int_J G_J(x, y) g(y)\, m(dy)$$

$$\eta(x) = -\int_J G_J(x, y) h(y)\, m(dy).$$

By (71),

$$f(x) = \lambda(x) + \gamma(x) - \eta(x).$$

Define μ by $\mu(dy) = g(y)\, m(dy)$ on (a, b). As (67a) implies, μ satisfies (56). And γ satisfies (59) with $\alpha = \beta = 0$. As (58a) implies, γ satisfies (55). So γ^+ exists, is finite, continuous from the right, and BV(loc), on (a, b); and $d\gamma^+ = g\, dm$. Similarly for η; and $d\eta^+ = h\, dm$. By inspection, λ^+ is finite, continuous from the right, and BV(loc), on (a, b); and $d\lambda^+ = 0$. Because everything is finite, on (a, b):

$$f^+ = \lambda^+ + \gamma^+ - \eta^+;$$

$$df^+ = d\lambda^+ + d\gamma^+ - d\eta^+ = (g - h)\, dm = \Gamma f\, dm.$$ ★

(73) Lemma. *If $m(a, a + \varepsilon) < \infty$ for ε with $0 < \varepsilon < b - a$, then f^+ is finite and continuous from the right even at a, for $f \in \Delta$.*

PROOF. Use (60a) twice, as for (72). ★

If $m(a, a + \varepsilon) = \infty$, nothing is asserted about the existence or continuity of f^+ at a.

(74) Proposition. *Let $f \in \Delta$. Then f^- exists, is finite, continuous from the left, and BV(loc), on (a, b). Moreover, $df^+ = df^-$ on (a, b). And $f^+ = f^-$ on (a, b) except maybe at x with $m\{x\} > 0$. If $m(b - \varepsilon, b) < \infty$ for ε with $0 < \varepsilon < b - a$, then f^- is finite and continuous from the left even at b.*

PROOF. As for (72, 73). ★

7. INFINITESIMAL GENERATOR

Recall assumption (61) that $\{P_x\}$ is regular and in natural scale on I. In this section, the infinitesimal generator (Γ, Δ) of $\{P_x\}$, as defined in (45), will be computed. This depends on I; the cases $I = (-\infty, \infty)$, $(0, \infty)$, $[0, \infty)$ are discussed. Any halfline can be converted to the positive (or nonnegative) halfline by translation and reflection. The case of bounded I is omitted; the results and arguments are very close to the halfline situation. Recall that m is the speed measure of $\{P_x\}$, while τ_x is the time to hit x, and $\tau_{[a,b]} = \tau_{ab}$ is the time to hit $\{a, b\}$, and $e_{ab}(x) = E_x\{\tau_{ab}\}$ for $x \in [a, b]$. As usual, E_x is expectation with respect to P_x.

The case $I = (-\infty, \infty)$

Suppose for (75) that $I = (-\infty, \infty)$. Let Δ_0 be the set of $f \in BC(-\infty, \infty)$ such that: f^+ exists, is finite, continuous from the right, and BV(loc), on $(-\infty, \infty)$: and $df^+ = g\,dm$ for some $g \in BC(-\infty, \infty)$. Plainly, Δ_0 is linear and g is unique. Define Γ_0 on Δ_0 by: $\Gamma_0 f = g$. Clearly, Γ_0 is a linear mapping of Δ_0 to $BC(-\infty, \infty)$. Informally, Δ_0 is the set of functions which have bounded, continuous second derivatives: where the first differentiation is with respect to Lebesgue measure, and the second with respect to speed measure. And Γ_0 is the second derivative operator.

(75) Theorem. $\Delta_0 = \Delta$ *and* $\Gamma_0 = \Gamma$.

PROOF. $\Delta \subset \Delta_0$ and $\Gamma_0 = \Gamma$ on Δ by (72). I plan to use (54). Suppose $f \in \Delta_0$ and $\Gamma_0 f \equiv f$. I need to show $f \equiv 0$. Suppose not. Find x_0 with $f(x_0) \neq 0$. Replace f by $-f$ if necessary to get $f(x_0) > 0$.

Suppose $f^+(x_0) \geqq 0$. Look at Figure 13. Let

$$z = \inf\{x : x > x_0 \text{ and } f(x) = 0\},$$

where the inf of an empty set is ∞. So $z > x_0$; the first problem is to show $z = \infty$. Confine x to (x_0, z). Remember $df^+ = f\,dm$ by assumption. Now

$$\text{(a)} \qquad f^+(x) = f^+(x_0) + \int_{(x_0, x]} f(y)\, m(dy)$$

is strictly increasing with x, because $f > 0$. So $f^+(x) > 0$. By (3.69),

$$\text{(b)} \qquad f(x) = f(x_0) + \int_{x_0}^{x} f^+(y)\, dy.$$

So $f(x)$ is strictly increasing with x. This forces $z = \infty$. Now f^+ is strictly increasing on (x_0, ∞) by (a). So $f^+(x) \geqq f^+(x_1) > 0$ when $x \geqq x_1 > 0$. Now (b) makes f unbounded on (x_0, ∞). This contradiction shows $f \equiv 0$.

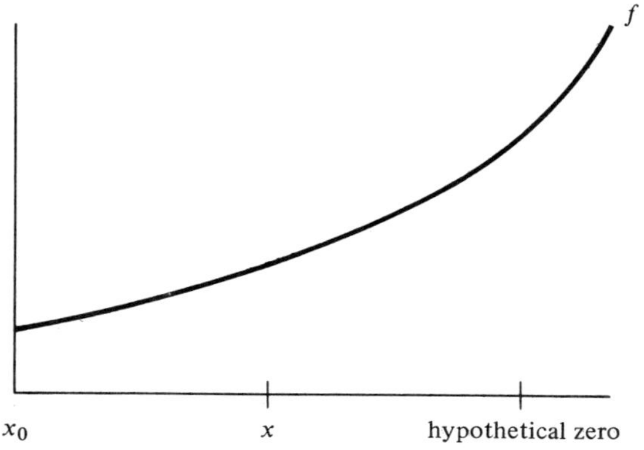

Figure 13.

Suppose $f^+(x_0) < 0$. Let $x < x_0$. Then

$$f^+(x_0) - f^+(x) = \int_{(x,x_0]} f(y)\, m(dy),$$

which increases strictly as x decreases until the first zero if any of f to the left of x_0, as in Figure 14. So $f^+(x) < 0$ and f increases strictly as x decreases until this zero, and this zero also disappears. So $f^+(x)$ is strictly decreasing with x, and f is unbounded. This contradiction shows $f \equiv 0$.

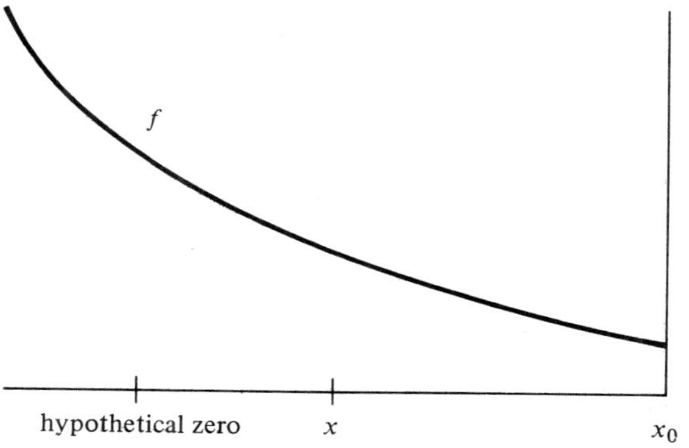

Figure 14.

Now use (54). ★

The case $I = (0, \infty)$

Suppose $I = (0, \infty)$ for (76) through (79).

(76) Lemma. **(a)** $P_u\{\text{hit } v\} = u/v$ *for* $0 < u \leqq v$.

(b) $P_u\{\text{hit } v\} = 1$ *for* $u \geqq v > 0$.

PROOF. *Claim* (a). Let $0 < u_n < u \leqq v$, and let $u_n \downarrow 0$. Then

$$\{X(0) = u \text{ and hit } v \text{ before } u_n\} \uparrow \{X(0) = u \text{ and hit } v\}.$$

Use (61):

$$P_u\{\text{hit } v\} = \lim_n P_u\{\text{hit } v \text{ before } u_n\} = \lim_n \frac{u - u_n}{v - u_n} = \frac{u}{v}.$$

Claim (b). Let $u_n > u \geqq v$ and $u_n \uparrow \infty$. Then

$$\{X(0) = u \text{ and hit } v \text{ before } u_n\} \uparrow \{X(0) = u \text{ and hit } v\}.$$

Use (61):

$$P_u\{\text{hit } v\} = \lim_n P_u\{\text{hit } v \text{ before } u_n\} = \lim_n \frac{u_n - u}{u_n - v} = 1.$$ ★

(77) Lemma. $P_x\{X(t) \to 0 \text{ as } t \to \infty\} = 1$ *for* $x > 0$.

PROOF. Fix ε and δ with $0 < \varepsilon < \delta < x$ and ε/δ small. Abbreviate $\sigma = \tau_\varepsilon$. Check that

$$\{\limsup_{t\to\infty} X(t) \leqq \delta\} \supset \{\sigma < \infty\} \cap \{S_\sigma \notin [\text{hit } \delta]\}.$$

By strong Markov (1c) and (76),

$$P_x\{\limsup_{t\to\infty} X(t) \leqq \delta\} \geqq P_x\{\sigma < \infty\}\cdot\{1 - P_\varepsilon[\text{hit } \delta]\}$$
$$= 1 - \frac{\varepsilon}{\delta}.$$

Let $\varepsilon \to 0$ and then let $\delta \downarrow 0$ through a sequence. ★

(78) Lemma. $\int_0^\varepsilon y.m(dy) = \infty$.

PROOF. Clearly, $\lim_{\delta\downarrow 0} \tau_{[\delta,2]} = \infty$ on $\{X(0) = 1\} \setminus \{\text{hit } 2\}$. But (77) makes $P_1\{\text{hit } 2\} = \frac{1}{2}$. So $\lim_{\delta\downarrow 0} E_1\{\tau_{[\delta,2]}\} = \infty$. By (67b),

$$E_1\{\tau_{[\delta,2]}\} = \frac{1-\delta}{2-\delta}\int_1^2 (2 - y)\, m(dy) + \frac{1}{2-\delta}\int_\delta^1 (y - \delta)\, m(dy).$$ ★

For (79), keep $I = (0, \infty)$. Let Δ_0 be the set of $f \in BC(0, \infty)$ such that: f^+ exists, is finite, continuous from the right, and $BV(\text{loc})$, on $(0, \infty)$; and $df^+ = g\, dm$ on $(0, \infty)$ for some $g \in BC(0, \infty)$. Let $\Gamma_0 f = g$. Plainly, Γ_0 is a well-defined linear mapping from the linear set Δ_0 to $BC(0, \infty)$.

(79) Theorem. $\Delta_0 = \Delta$ *and* $\Gamma_0 = \Gamma$.

PROOF. $\Delta \subset \Delta_0$ and $\Gamma_0 = \Gamma$ on Δ by (72). I plan to use (54). Suppose $f \in \Delta_0$ and $\Gamma_0 f = f$. Suppose $f(x_0) \neq 0$. Replace f by $-f$ if need be to get $f(x_0) > 0$. Suppose $f^+(x_0) < 0$; the opposite case proceeds as in the proof for (75). Suppose $0 < x < x_0$ and $f > 0$ on (x, x_0). Then

$$f^+(x_0) - f^+(x) = \int_{(x,x_0]} f\,dm > 0,$$

so $f^+(x) < f^+(x_0) < 0$. Consequently f is positive and strictly decreasing on $(0, x_0)$, as in Figure 15. But f is bounded at 0 by the definition of Δ_0;

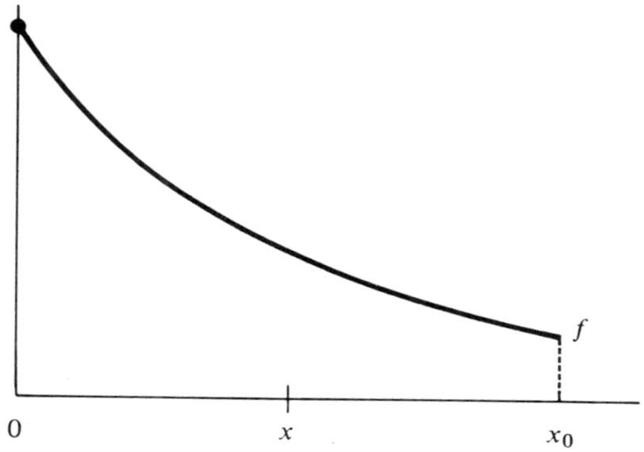

Figure 15.

define $f(0) = F(0+)$. Then F satisfies (55) with $a = 0$ and $b = x_0$ and $d\mu = f\,dm$. So (58b) contradicts (78). This proves $f \equiv 0$, and (54) applies. ★

The case $I = [0, \infty)$ and 0 is absorbing

From (80) and (81), suppose $I = [0, \infty)$ and 0 is absorbing.

(80) Lemma. **(a)** $\int_0^\varepsilon y\,m(dy) < \infty$.

(b) $P_x\{X(t) = 0$ *for all large* $t\} = 1$.

PROOF. *Claim* (a) follows from (67a), because $[0, \varepsilon] \subset I$.

Claim (b). Let $x_n > x$ and $x_n \uparrow \infty$. Then

$$\{X(0) = x \text{ and hit 0 before } x_n\} \uparrow \{X(0) = x \text{ and hit } 0\};$$

with the help of (61),

$$P_x\{\text{hit } 0\} = \lim_n P_x\{\text{hit } 0 \text{ before } x_n\} = \lim_n \frac{x_n - x}{x_n} = 1.$$

By (6) and strong Markov (1c),

$$P_x\{X(t) = 0 \text{ for all large } t\} = P_x\{\text{hit } 0\}. \qquad \bigstar$$

Let Δ_0 be the set of $f \in BC[0, \infty)$ such that: f^+ exists, is finite, continuous from the right, and $BV(\text{loc})$, on $(0, \infty)$; and $df^+ = g\,dm$ on $(0, \infty)$ for some $g \in BC[0, \infty)$ with $g(0) = 0$. Let $\Gamma_0 f = g$. Plainly, Γ_0 is a well-defined linear mapping of the linear set Δ_0 into $BC[0, \infty)$.

(81) Theorem. $\Delta = \Delta_0$ *and* $\Gamma = \Gamma_0$.

PROOF. As in (75). Use (52) and (72) to check $\Delta \subset \Delta_0$ and $\Gamma_0 = \Gamma$ on Δ. Suppose $f \in \Delta_0$ and $\Gamma_0 f = f$. Suppose $f(x_0) \neq 0$ for some $x_0 \in (0, \infty)$. Replace f by $-f$ if necessary to get $f(x_0) > 0$. Suppose $f^+(x_0) < 0$, the other case being familiar. Then f is strictly decreasing on $[0, x_0]$. Since f vanishes at 0, a contradiction is reached. This proves $f \equiv 0$. Now you can use (54). ★

The case $I = [0, \infty)$ and 0 is instantaneous

Suppose for (82–89) that $I = [0, \infty)$ and 0 is instantaneous. Let $e(x) = E_0(\tau_x)$, which is finite by (20).

(82) Lemma. **(a)** *e is nondecreasing, continuous, and strictly convex on $(0, \infty)$.*
(b) *$de^+ = dm$ on $(0, \infty)$.*
(c) *$e^+(0)$ exists, is nonnegative, and finite.*
(d) *$m(0, \varepsilon) < \infty$.*

PROOF. Let $0 \leqq x \leqq b$, as in Figure 16.

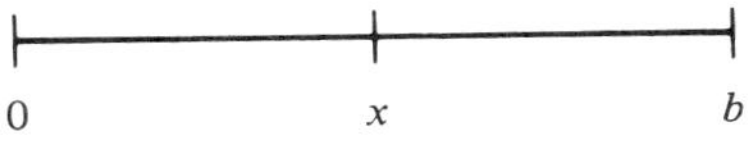

Figure 16.

Using (61) and the strong Markov property (1d), as for (63),

(83) $$e(b) = e(x) + E_x\{\tau_b\}$$

(84) $$e(b) = e(x) + e_{0b}(x) + \frac{b-x}{b}e(b);$$

so

(85) $$e(x) = \frac{x}{b}e(b) - e_{0b}(x).$$

Relation (83) shows e is nondecreasing; (62) and relation (85) complete the proof of (a). Differentiate (85) and use (66) to get (b). Assertion (c) follows from (a) and (3.74h). Assertion (d) follows from (60a) applied to $-f$. Condition (56) holds by (67a). ★

(86) Definition. *$m(0) = e^+(0)$ is the derivative of $e(x) = E_0\{\tau_x\}$ at $x = 0$. If $m(0) > 0$, then 0 is a* slow reflector. *If $m(0) = 0$, then 0 is a* fast reflector.

(87) Fact. $e(x) = E_0\{\tau_x\} = \int_0^x m[0, z]\,dz.$

PROOF. By (82b), for $0 < y < z$,

$$e^+(z) = e^+(y) + m(y, z].$$

Lemma (82a) and (3.74h) make e^+ continuous at 0. Let $y \downarrow 0$, and remember $e^+(0) = m(0)$ by definition (87).

(88) $$e^+(z) = e^+(0) + m(0, z] = m[0, z].$$

By Dini (3.69),

$$e(x) = e(0) + \int_0^x e^+(z)\,dz;$$

but $e(0) = 0$. ★

Let Δ_0 be the set of $f \in BC[0, \infty)$ such that: f^+ exists, is finite, and continuous from the right on $[0, \infty)$; and f^+ is BV(loc) on $(0, \infty)$, with $df^+ = g\,dm$ on $(0, \infty)$ for some $g \in BC[0, \infty)$; and $f^+(0) = m(0)g(0)$. Let $\Gamma_0 f = g$. Clearly, Γ_0 is a well-defined linear mapping of the linear set Δ_0 into $BC[0, \infty)$.

(89) Theorem. $\Delta_0 = \Delta$ *and* $\Gamma_0 = \Gamma$.

PROOF. As in (75). The first step is to show (Γ_0, Δ_0) extends (Γ, Δ). This follows from (72), and a separate argument to take care of 0. Namely, suppose $f \in \Delta$. Then f^+ exists, finite and continuous, even at 0, by (72) and (73). The condition of (73) is satisfied by (82). By (51),

$$\Gamma f \cdot 0 = \lim_{\varepsilon \downarrow 0} \frac{f(\varepsilon) - f(0)}{e(\varepsilon)}.$$

If $m(0) = 0$, then $e(\varepsilon) = o(\varepsilon)$. But $\Gamma f \cdot 0 < \infty$, so $f^+(0) = 0$. There is nothing else to say about $\Gamma f \cdot 0$ in this case. If $m(0) > 0$, then $e(\varepsilon) = \varepsilon m(0) + o(\varepsilon)$, and $f^+(0)/m(0) = \Gamma f \cdot 0$. This completes the proof that (Γ_0, Δ_0) extends (Γ, Δ).

Suppose $f \in \Delta_0$ and $\Gamma_0 f = f$. I have to show $f \equiv 0$. Suppose not. Suppose $f(x_0) > 0$ and $f^+(x_0) < 0$ for some $x_0 \in (0, \infty)$. The other cases are routine. As usual, f^+ is strictly decreasing on $[0, x_0]$, so $f^+(0) < 0$, a contradiction. See Figure 18. This proves $f \equiv 0$, and (54) applies.

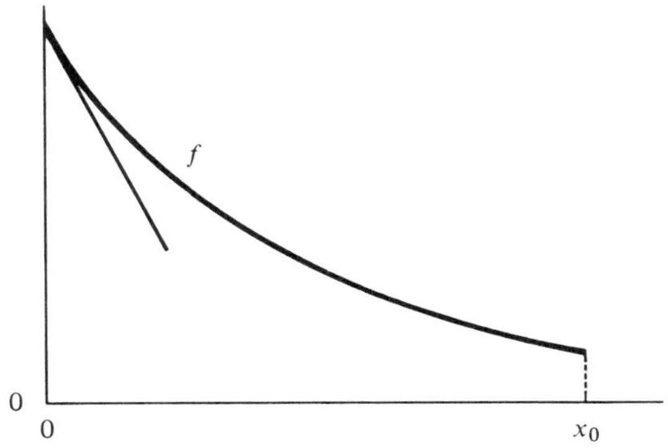

Figure 18. ★

General I

(90) Theorem. *For any I, the diffusion $\{P_x\}$ is determined by its speed measure, with the convention $m\{x\} = \infty$ for absorbing x.*

PROOF. If $I = (-\infty, \infty)$ or $(0, \infty)$ or $[0, \infty)$, use (75 or 79 or 81 or 89) to see that m determines (Γ, Δ). Then use (47). For other I, improvise. ★

Brownian motion

(91) Theorem. *Standard Brownian motion is a regular diffusion in natural scale on $(-\infty, \infty)$. Its speed measure is twice Lebesgue measure; Δ is the set of $f \in BC(-\infty, \infty)$ with $\ddot{f} \in BC(-\infty, \infty)$; and $\Gamma f = \frac{1}{2}\ddot{f}$.*

Here $\ddot{f}$ is the second derivative of f. If $f \in \Delta$, then $\dot{f} \in BC(-\infty, \infty)$ by calculus: $\dot{f}$ is the first derivative. Remember (2).

PROOF. Using (1.114), verify that standard Brownian motion is in natural scale, and $e_{ab}(x) = (x - a)(b - x)$ for $a \leqq x \leqq b$. Differentiate twice to get $m(dx) = 2\,dx$. Suppose $f \in \Delta$; let $g = \Gamma f \in BC(-\infty, \infty)$. As (72) implies: f^+ exists, is continuous from the right, and is $BV(\text{loc})$ on $(-\infty, \infty)$; moreover,

$$f^+(ds) = 2g(s)\,ds \quad \text{on } (-\infty, \infty).$$

That is,

$$f^+(y) - f^+(x) = 2\int_x^y g(s)\,ds.$$

In particular, f^+ is continuous; and is therefore the usual derivative of f by (3.67). By calculus, $\ddot{f} = 2g \in BC(-\infty, \infty)$, and

$$\Gamma f = g = \tfrac{1}{2}\ddot{f}.$$

Conversely, let $f \in BC(-\infty, \infty)$ with $g = \frac{1}{2}\ddot{f} \in BC(-\infty, \infty)$. Then f^+ exists, is continuous from the right, and is $BV(\text{loc})$ on $(-\infty, \infty)$; by calculus,

$$f^+(y) - f^+(x) = \int_x^y g(s)\,2\,ds.$$

So $f \in \Delta$ by (75). ★

8. BROWNIAN LOCAL TIME

Let $\{B_x\}$ be standard Brownian motion, as defined for (2). Briefly, Ω is the set of all continuous functions from $[0, \infty)$ to $(-\infty, \infty)$, and B_x is a probability on Ω. The coordinate process X on Ω is Brownian motion starting from x, relative to B_x. Write E_x for expectation with respect to B_x. Let J be a proper, closed, finite interval $[a, b]$. Recall that τ_J is the least t if any with $X(t) = a$ or b, and $\tau_J = \infty$ if none. Remember the Green's function G_J from (57). Let f be a continuous function on $(-\infty, \infty)$, and let $F(t) = \int_0^t f[X(s)]\,ds$.

(92) Lemma. $E_x[F(\tau_J)] = 2\int_J G_J(x, y) f(y)\,dy$ *for* $x \in J$.

PROOF. Use (68) and (91) for bounded f. Then check that both sides of (92) depends only on f retracted to J. ★

Suppose $f \geqq 0$ and $\{x : f(x) > 0\}$ is dense. Let G be the set of ω such that $F(0, \omega) = 0$, while $F(\cdot, \omega)$ is continuous and strictly increasing, and $F(\infty, \omega) = \infty$.

(93) Lemma. $B_x\{G\} = 1$ *for all* x.

PROOF. You can check that $F(\cdot, \omega)$ is continuous and nondecreasing and $F(0, \omega) = 0$, for all ω. For the rest, let $\sigma_0 = 0$. Let σ_{n+1} be the least $t \geqq \sigma_n + 1$ with $X(t) = 0$. Let

$$Z_n = \int_{\sigma_n}^{\sigma_{n+1}} f[X(t)]\,dt.$$

As strong Markov (1.21) implies: $Z_0, Z_1, \ldots$ are B_x-independent; for $n \geqq 1$, the B_x-distribution of Z_n depends neither on n nor on x, and coincides with the B_0-distribution of Z_0. Use (1.48), the density of $\{f > 0\}$, and continuity: $\{f \circ \omega > 0\}$ is open and dense for B_x-almost all ω. So

$$B_0\{Z_0 > 0\} = 1 \qquad \text{and} \qquad B_x\{F(s) < F(t)\} = 1 \quad \text{for } s > t.$$

In particular,

$$B_x\{F(\infty) = \Sigma_n Z_n = \infty\} = 1.$$

Finally, F is strictly increasing iff $F(s) < F(t)$ for all pairs of nonnegative rationals s and t with $s < t$. ★

On a first reading, the balance of this section can be skipped. For $\omega \in \Omega$ and $t \geqq 0$, let $\mu(t, \omega)(\cdot)$ be the image of Lebesgue measure on $[0, t]$ under the mapping $s \to \omega(s)$. More explicitly, for Borel A,

$$\mu(t, \omega)\{A\} = \text{Lebesgue}\,\{s : 0 \leqq s \leqq t \text{ and } \omega(s) \in A\}.$$

Thus, $\mu(t, \omega)\{\cdot\}$ is a measure on the Borel subsets of the line, having total mass t. Abbreviate $\mu(t, \omega, A) = \mu(t, \omega)\{A\}$. The next result is due to (Trotter, 1958). The proof is difficult and is omitted.

(94) Theorem. *For each $\omega \in \Omega$, there is a continuous nonnegative function $(x, t) \to \lambda(x, t, \omega)$, such that $\omega \to \lambda(x, t, \omega)$ is measurable, and $B_x\{\Omega_0\} = 1$ for all x, where Ω_0 is the set of $\omega \in \Omega$ with*

$$\mu(t, \omega, A) = \int_A \lambda(y, t, \omega)\, dy$$

for all Borel A and $t \geqq 0$.

NOTE. Ω_0 is measurable, because A can be confined to the set of rational intervals, and t to the set of rational times.

(95) Remark. For each $\omega \in \Omega_0$: for all y and t,

$$\lambda(y, t, \omega) = \lim \frac{\mu(t, \omega)[y - \varepsilon, y + \delta]}{\varepsilon + \delta}$$

as $\varepsilon \to 0$ and $\delta \to 0$ with $\varepsilon \geqq 0$ and $\delta \geqq 0$ and $\varepsilon + \delta > 0$.

NOTE. The set of ω such that the limit in (95) exists for all y and t, and defines a continuous function of y and t, is measurable. To see this, use the technique of (Dubins and Freedman, 1964, Theorem 4.4). This set of ω is the largest possible Ω_0.

INFORMAL NOTE. λ is called *local time*: $\lambda(y, t)\, dy$ is the time Brownian motion spends in $(y, y + dy)$, up to time t.

WARNING. There is a mystery constant $\frac{1}{2}$ in this theory; it's the same $\frac{1}{2}$ which occurs in the $N(0, 1)$ density. My local time is twice other people's local time, but I get the $\frac{1}{2}$ back in (113).

Remember

$$(S_t\omega)(s) = \omega(t + s), \text{ and } J = [a, b].$$

(96) Lemma. **(a)** *For all s, t, ω, A:*

$$\mu(t + s, \omega, A) = \mu(t, \omega, A) + \mu(s, S_t\omega, A).$$

For (b–g), confine ω to Ω_0.

(b) *For all s, t, y:*

$$\lambda(y, t + s, \omega) = \lambda(y, t, \omega) + \lambda(y, s, S_t\omega).$$

(c) $\lambda(y, 0, \omega) = 0$.

(d) *$\lambda(y, \cdot, \omega)$ is nondecreasing.*

Define $\lambda(y, \infty, \omega) = \lim_{t\to\infty} \lambda(y, t, \omega)$.

(e) *Suppose $\omega(0) \in J$. Then $\lambda[y, \tau_J(\omega), \omega]$ is nondecreasing with J.*

(f) *Suppose* $\omega(0) \in J$. *If* $y \leqq a$ *or* $y \geqq b$, *then* $\lambda[y, \tau_J(\omega), \omega] = 0$.

(g) $\lambda[y, \tau_y(\omega), \omega] = 0$.

PROOF. *Claim* (*a*) is easy.

Claim (*b*) follows from (a) and (95).

Claim (*c*) follows from (95).

Claim (*d*) follows from (b).

Claim (*e*) follows from (d), because $\tau_J(\omega)$ is nondecreasing with J, on $\{X(0) \in J\}$.

Claim (*f*). Suppose $\tau_J(\omega) < \infty$, the other case being easier. If $y < a$ or $y > b$, use (95) to get $\lambda[y, \tau_J(\omega), \omega] = 0$. Finally, use continuity to handle $y = a$ or b.

Claim (*g*) is like (f). ★

As usual, $\lambda(y, \tau_J)$ is the random variable

$$\omega \to \lambda[y, \tau_J(\omega), \omega].$$

(97) Theorem. *For* x *and* y *in* J,

$$E_x[\lambda(y, \tau_J)] = 2G_J(x, y). \tag{98}$$

The proof is deferred; the main effort is devoted to showing that the left side of (98) is continuous. This is accomplished by (99–104). For $c > 0$, let $cJ = [ac, bc]$. For $y \in J$, let $J - y = [a - y, b - y]$. Keep x and y in J.

(99) Lemma. **(a)** *The* B_x-*distribution of* $\lambda(y, \tau_J)$ *coincides with the* B_{x-y}-*distribution of* $\lambda(0, \tau_{J-y})$.

(b) *The* B_{cx}-*distribution of* $\lambda(cy, \tau_{cJ})$ *coincides with the* B_x-*distribution of* $c\lambda(y, \tau_J)$.

PROOF. *Claim* (*a*) is similar to (b).

Claim (*b*). Use the scaling property (1.11b) to see that the B_{cx}-distribution of

$$c^{-2}\mu(\tau_{cJ}, \cdot)[cy - c\varepsilon, cy + c\delta]$$

coincides with the B_x-distribution of

$$\mu(\tau_J, \cdot)[y - \varepsilon, y + \delta].$$

Divide by $\varepsilon + \delta$ and use (95). ★

Let δ_0 be point mass at 0. Remember $J = [a, b]$.

(100) Lemma. **(a)** *For $a < x \leqq y < b$, the B_x-distribution of $\lambda(y, \tau_J)$ is*

$$\frac{x-a}{y-a} \cdot [B_y\text{-distribution of } \lambda(y, \tau_J)] + \frac{y-x}{y-a} \cdot \delta_0.$$

(b) *For $a < y \leqq x < b$, the B_x-distribution of $\lambda(y, \tau_J)$ is*

$$\frac{b-x}{b-y} \cdot [B_y\text{-distribution of } \lambda(y, \tau_J)] + \frac{x-y}{b-y} \cdot \delta_0.$$

PROOF. *Claim* (*a*). Suppose $x < y$. Confine ω to $\{X(0) = x \text{ and } \Omega_0\}$, which has B_x-probability 1. Abbreviate $\sigma = \tau_y$. I say

$$\begin{aligned} \lambda(y, \tau_J) &= 0 && \text{on } \{\text{hit } a \text{ before } y\} \\ &= \lambda(y, \tau_J) \circ S_\sigma && \text{on } \{\text{hit } y \text{ before } a\}. \end{aligned}$$

For the first equality, use the path continuity to get

$$\tau_J \leqq \sigma \qquad \text{on } \{\text{hit } a \text{ before } y\};$$

then use (96d, g). For the second equality, check

$$\tau_J = \sigma + \tau_J \circ S_\sigma \quad \text{on } \{\text{hit } y \text{ before } a\};$$

then use (96b) with $t = \sigma$ and $s = \tau_J \circ S_\sigma$; finally, use (96g) to eliminate the first term in (96b). Let ϕ be a bounded, measurable function on $(-\infty, \infty)$. Abbreviate

$$\begin{aligned} A &= \{\text{hit } a \text{ before } y\} \\ C &= \{\text{hit } y \text{ before } a\}. \end{aligned}$$

Now $A \cup C$ is B_x-almost all of Ω by (1.17). So

$$E_x\{\phi[\lambda(y, \tau_J)]\} = \int_A \phi[\lambda(y, \tau_J)]\, dB_x + \int_C \phi[\lambda(y, \tau_J)]\, dB_x.$$

Check $B \in \mathscr{B}(\sigma+)$. Use (1.14), and strong Markov in the second computation:

$$\begin{aligned} \int_A \phi[\lambda(y, \tau_J)]\, dP_x &= \int_A \phi(0)\, dB_x \\ &= B_x\{A\} \circ \phi(0) \\ &= \frac{y-x}{y-a} \cdot \phi(0); \end{aligned}$$

$$\int_C \phi[\lambda(y, \tau_J)]\, dP_x = \int_C \phi[\lambda(y, \tau_J) \circ S_\sigma]\, dB_x$$
$$= B_x\{C\} \cdot E_y\{\phi[\lambda(y, \tau_J)]\}$$
$$= \frac{x - a}{y - a} \cdot E_y\{\phi[\lambda(y, \tau_J)]\}.$$

Claim (*b*) is symmetric. ★

(101) Lemma. *For* $0 \leqq f < \infty$, *let* $\theta(f) = E_0\{\lambda(0, \tau_{[-1, f]})\}$. *For* (*a*) *and* (*b*), *let* $f = (b - y)/(y - a)$. *Then*

$$\textbf{(a)} \quad E_x\{\lambda(y, \tau_J)\} = (x - a)\theta(f) \qquad \textit{for } a < x \leqq y < b$$

while

$$\textbf{(b)} \quad E_x\{\lambda(y, \tau_J)\} = \frac{b - x}{b - y}(y - a)\theta(f) \textit{ for } a < y \leqq x < b.$$

PROOF. I will only argue the first claim.

$$E_x\{\lambda(y, \tau_J)\} = \frac{x - a}{y - a} \cdot E_y\{\lambda(y, \tau_J)\} \qquad \text{by (100a)}$$
$$= \frac{x - a}{y - a} \cdot E_0\{\lambda(0, \tau_{J - y})\} \qquad \text{by (99a)}$$
$$= (x - a) \cdot E_0\{\lambda(0, \tau_{[-1, f]})\} \quad \text{by (99b).}$$ ★

(102) Lemma. *Fix* $x \in J$. *Then*

$$E_x[\lambda(y, \tau_J)] = 2G_J(x, y)$$

for Lebesgue almost all $y \in J$.

PROOF. Let $f \in BC(-\infty, \infty)$. Let

$$F(t) = \int_0^t f[X(s)]\, ds$$
$$= \int_{-\infty}^{\infty} f(y)\lambda(y, t)\, dy \quad \text{on } \Omega_0,$$

by change of variables. By (96f), on $\{\Omega_0 \text{ and } \omega(0) \in J\}$,

$$F(\tau_J) = \int_J f(y)\lambda(y, \tau_J)\, dy.$$

Suppose $f \geqq 0$. By Fubini and (92),

$$E_x\{F(\tau_J)\} = \int_J f(y) E_x\{\lambda(y, \tau_J)\}\, dy$$

$$= 2 \int_J f(y) G_J(x, y)\, dy.$$

The rest is routine. ★

(103) Lemma. *Fix $x \in J$. Then*

$$E_x\{\lambda(y, \tau_J)\} < \infty \quad \textit{for all } y \in J.$$

Proof. Suppose $a < x < b$ and $a < y < b$, the other cases being easy. Lemma (102) produces one $y \in (x, b)$ with $E_x\{\lambda(y, \tau_J)\} < \infty$. Let f be $(b - y)/(a - y)$. As (101a) implies, $\theta(f) < \infty$. By (99b), for any constant $c > 0$,

$$E_0\{\lambda(0, \tau_{[-c, cf]})\} < \infty.$$

Take c large and use (96d) to see $\theta(f) < \infty$ for all $f \in (0, \infty)$. Use (101) again. ★

(104) Lemma. *θ is finite and continuous.*

Proof. To begin with, θ is finite by (103). Let $0 \leqq g < f$, as in Figure 19.

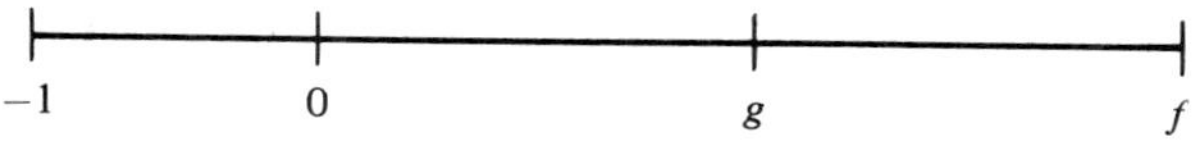

Figure 19.

I say

$$\theta(f) = \theta(g) + \frac{1}{1 + g} E_g\{\lambda(0, \tau_{[-1, f]})\}.$$

Temporarily, let $\gamma = \tau_{[-1, g]}$; let $\phi = \tau_{[-1, f]}$; let

$$G = \{\omega(0) = 0 \text{ and hit } g \text{ before } -1\}$$

$$H = \{\omega(0) = 0 \text{ and hit } -1 \text{ before } g\}.$$

Now

$$\phi = \gamma \qquad \text{on } H;$$

so

$$\lambda(0, \phi) = \lambda(0, \gamma) \qquad \text{on } H.$$

Moreover,

$$\phi = \gamma + \phi \circ S_\gamma \quad \text{on } G;$$

so (96b) makes

$$\lambda(0, \phi) = \lambda(0, \gamma) + \lambda(0, \phi) \circ S_\gamma \quad \text{on } G \cap \Omega_0$$

But G is in $\mathscr{B}(\gamma+)$. Use (94), (1.114) and strong Markov:

$$\begin{aligned}\theta(f) &= E_0\{\lambda(0, \phi)\}\\ &= \int_H \lambda(0, \phi)\, dB_0 + \int_G \lambda(0, \phi)\, dB_0\\ &= E_0\{\lambda(0, \gamma)\} + \int_G \lambda(0, \phi) \circ S_\gamma\, dB_0\\ &= E_0\{\lambda(0, \gamma)\} + B_0\{G\} \int \lambda(0, \phi)\, dB_g\\ &= \theta(g) + \frac{1}{1+g} \cdot E_g\{\lambda(0, \phi)\}.\end{aligned}$$

Use (101b) with g for x and 0 for y and -1 for a and f for b:

$$E_g\{\lambda(0, \tau_{[-1,f]})\} = \frac{f-g}{f} \cdot \theta(f).$$

So

$$\theta(f) = \theta(g) + \frac{1}{1+g} \cdot \frac{f-g}{f} \cdot \theta(f);$$

or,

$$\theta(g) = \frac{g(1+f)}{f(1+g)} \cdot \theta(f) \quad \text{for } f > g \geqq 0. \qquad \bigstar$$

Proof of (97). Suppose $x = a$ or b. The right side of (98) vanishes by definition (57). Confine ω to $\{\Omega_0$ and $X(0) = x\}$. Then $\tau_J = 0$, and $\lambda(y, \tau_J) = 0$ by (96c); so the left side of (98) vanishes. Suppose $y = a$ or b. The right side of (98) vanishes by definition (57). Confine ω to $\{\Omega_0$ and $X(0) = x\}$. Then $\lambda(y, \tau_J) = 0$ by (96f); so the left side of (98) vanishes.

Fix x with $a < x < b$. Combine (101) and (104) to see that $y \to E_x\{\lambda(y, \tau_J)\}$ is continuous on (a, b). Now use (102). ★

(105) Remark. Similar arguments, pressed harder, show:

(a) $\dfrac{1}{\varepsilon + \delta}\mu(t, \cdot)[y - \varepsilon, y + \delta]$ is uniformly B_0-integrable as $\varepsilon, \delta \to 0$ with $\varepsilon \geqq 0$ and $\delta \geqq 0$ and $\varepsilon + \delta > 0$, for all y and bounded t;

(b) $\lambda(0, \tau_J)$ is B_0-exponentially distributed when $0 \in J$.

The next main result is (119); here are some preliminaries.

(106) Lemma. **(a)** *For each x, the set of* ω *such that* $\lambda(x, \infty, \omega) = \infty$ *has* B_x*-probability* 1.

(b) *For each x, the set of* ω *such that* $\lambda(x, t, \omega) > 0$ *for all t has* B_x*-probability* 1.

(c) *For each x and y, the set of* ω *such that* $\lambda(y, \infty, \omega) = \infty$ *has* B_x*-probability* 1.

PROOF. It is enough to do the case $x = 0$.

Claim (a). Let $\varepsilon > 0$. Let $\sigma_{0,\varepsilon} = 0$. As in Figure 20, let $\theta_{n+1,\varepsilon}$ be the least $t > \sigma_{n,\varepsilon}$ if any with $X(t) = \varepsilon$, and $\theta_{n+1,\varepsilon} = \infty$ if none. For $n = 1, 2, \ldots,$ let $\sigma_{n,\varepsilon}$ be the least $t > \theta_{n,\varepsilon}$ if any with $X(t) = 0$, and $\sigma_{n,\varepsilon} = \infty$ if none. These quantities are Markov times, which are finite B_0-almost surely by (1.17).

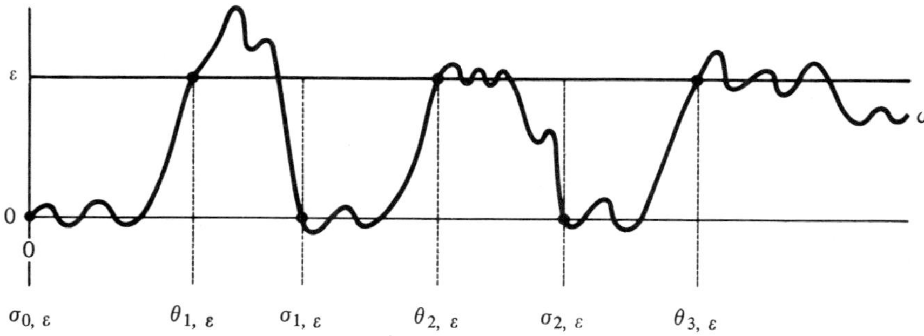

Figure 20.

Let $L_n = \lambda(0, \sigma_{n,\varepsilon}) - \lambda(0, \sigma_{n-1,\varepsilon})$. By telescoping,

(107) $$\lambda(0, \sigma_{n,\varepsilon}) = L_1 + \cdots + L_n.$$

Abbreviate S_n for the shift $S_{\sigma_{n,\varepsilon}}$. Check

(108a) $$\sigma_{n+1,\varepsilon} - \sigma_{n,\varepsilon} = \sigma_{1,\varepsilon} \circ S_n;$$

and with the help of (96b),

(108b) $$L_{n+1} = L_1 \circ S_n \quad \text{on } \Omega_0.$$

Use (94) and strong Markov (1.21):

(109a) $[(\sigma_{n+1,\varepsilon} - \sigma_{n,\varepsilon}), L_{n+1}]$ are B_0-independent and identically distributed for $n = 0, 1, \ldots$.

Using (94, 96d),

(109b) $$B_0\{L_n \geqq 0\} = 1.$$

Check $\sigma_{1,\varepsilon} \geqq \tau_{[-1,\varepsilon]}$ on $\{X(0) = 0\}$; so (96*d*) makes

$$L_1 = \lambda(0, \sigma_{1,\varepsilon}) \geqq \lambda(0, \tau_{[-1,\varepsilon]})$$

on Ω_0. Theorem (97) makes $E_0\{\lambda(0, \tau_{[-1,\varepsilon]})\} > 0$, so (94) makes

(110) $$B_0\{L_1 > 0\} > 0.$$

Therefore,

$$B_0\{\lambda(0, \sigma_{n,\varepsilon}) \to \infty\} = 1,$$

proving claim (a).

Claim (*b*). By the scaling property (1.11b),

$$\delta = B_0\{\lambda(0, \sigma_{1,\varepsilon}) > 0\}$$

does not depend on $\varepsilon > 0$. And $\delta > 0$ by (110). By (107, 109),

$$B_0\{\lambda(0, \sigma_{n,\varepsilon}) = 0\} = (1 - \delta)^n.$$

Fix $t > 0$. Then

$$B_0\{\lambda(0, t) = 0\} \leqq B_0\{\sigma_{n,\varepsilon} > t\} + (1 - \delta)^n.$$

Choose n so large that the second term is small. Then choose ε so small that the first term is small: this is feasible by (109a), because $\sigma_{1,\varepsilon} \to 0$ in B_0-probability as $\varepsilon \to 0$, by (1.57). Therefore,

$$B_0\{\lambda(0, t) = 0\} = 0.$$

Finally, $\lambda(0, t) > 0$ for all t iff $\lambda(0, t) > 0$ for all rational t.

Claim (*c*). As (91) shows, $B_x\{\tau_y < \infty\} = 1$. On $\{\Omega_0$ and $\tau_y < \infty\}$,

$$\{\lambda(y, \infty) = \infty\} = S_{\tau_y}^{-1}\{\lambda(y, \infty) = \infty\}$$

by (96b). By strong Markov (1.21) and (94),

$$B_x\{\lambda(y, \infty) = \infty\} = B_y\{\lambda(y, \infty) = \infty\}.$$

Use (a). ★

As before, $S_y(\omega) = \{t : \omega(t) = y\}$. The *support* of a measure on $(-\infty, \infty)$ is the complement of its largest open null set. If F is a nondecreasing function on $(-\infty, \infty)$, then $F(dx)$ is the measure whose distribution function is F, and the support of F is the support of $F(dx)$.

(111) Lemma. *For $\omega \in \Omega_0$, the support of $\lambda(y, dt, \omega)$ is included in $S_y(\omega)$.*

PROOF. If $[a, b]$ is a positive distance from $S_y(\omega)$, then

$$|\omega(t) - y| \geqq d = d(\omega) > 0 \quad \text{for } a \leqq t \leqq b.$$

And (95) makes $\lambda(y, a, \omega) = \lambda(y, b, \omega)$. ★

(112) Lemma. **(a)** *The set of pairs (y, ω) such that $\omega \in \Omega_0$ and $\lambda(y, dt, \omega)$ has support $S_y(\omega)$ is product measurable.*

(b) *For each y, for B_x-almost all ω, the measure $\lambda(y, dt, \omega)$ has $S_y(\omega)$ for support.*

PROOF. *Claim (a).* By (111), I only have to show that the set G of (y, ω) with $\omega \in \Omega_0$ and

$$\text{support } \lambda(y, dt, \omega) \supset S_y(\omega)$$

is product measurable. For nonnegative rational r, let $\tau(r, y)$ be the least $t \geqq \tau$ if any with $X(t) = y$, and $\tau(r, y) = \infty$ if none. I say that $(y, \omega) \to \tau(r, y, \omega)$ is product measurable. In fact, $\tau(r, y, \omega) \leqq s$ iff

$$0 = \inf_t \{|\omega(t) - y| : r \leqq t \leqq s \text{ and } t \text{ is rational}\}.$$

And G is the set of (y, ω) with $\omega \in \Omega_0$ and: either $\tau(r, y, \omega) = \infty$, or

$$\lambda[y, \tau(r, y, \omega) + s, \omega] > \lambda[y, \tau(r, y, \omega), \omega]$$

for all nonnegative rational r and positive rational s.

Claim (b). Continue with the argument for (a). Fix x, y, s, r. Abbreviate $\sigma = \tau(r, y)$. Let H be the set of $\omega \in \Omega_0$ with $\sigma(\omega) < \infty$ and

$$\lambda[y, \sigma(\omega) + s, \omega] > \lambda[y, \sigma(\omega), \omega].$$

As (96b) implies, H is also the set of $\omega \in \Omega_0$ with $\sigma(\omega) < \infty$ and

$$S_\sigma(\omega) \in \{\lambda(y, s) > 0\}.$$

But σ is Markov, and $B_x\{\sigma < \infty\} = 1$ by (1.17). Check $X(\sigma) = y$ on $\{\sigma < \infty\}$. Remember $B_x\{\Omega_0\} = 1$ by (94). Use strong Markov and (106b):

$$B_x\{H\} = B_y\{\lambda(y, s) > 0\} = 1.$$ ★

Let m be a nontrivial measure on $(-\infty, \infty)$. Let K be the support of m; that is, K is the complement of the largest open set with m-measure 0. Let $a = \inf K$ and $b = \sup K$. Suppose m is locally finite on (a, b), while $m\{a\}$ and $m\{b\}$ are finite. If $a = -\infty$, then $m\{a\} = 0$; similarly for b. Let $\dot{m}$ be the density of m with respect to Lebesgue measure, when m is absolutely continuous. Let

(113) $$M(t) = \frac{1}{2}\int_0^t \dot{m}[X(s)]\, ds = \frac{1}{2}\int_{-\infty}^{\infty} \lambda(y, t)\, m(dy).$$

Think about the middle expression only for absolutely continuous m. Use (96b) to check that for $\omega \in \Omega_0$:

(114) $$M(t + s, \omega) = M(t, \omega) + M(s, S_t\omega) \quad \text{for all } s, t.$$

If F is a nondecreasing function on $(-\infty, \infty)$, let $|F|$ or $F(dx)$ be the measure whose distribution function is F, and let the support of F be the support of $|F|$. Let

$$S_x(\omega) = \{t : t \geqq 0 \text{ and } \omega(t) = x\}.$$

Let

$$S_A(\omega) = \{t : t \geqq 0 \text{ and } \omega(t) \in A\}.$$

As (96) implies: for $\omega \in \Omega_0$,

(115) $M(0) = 0$, and $0 \leqq M(t) \leqq \infty$, and M is nondecreasing.

Let $\varphi(\omega) = \sup\{t : t \geqq 0 \text{ and } M(t, \omega) < \infty\}$. Thus, $M(t, \omega)$ is finite for $t < \varphi(\omega)$ and infinite for $t > \varphi(\omega)$. However, $M[\varphi(\omega), \omega]$ can be finite or infinite; the first possibility is unlikely. If $\omega \in \Omega_0$, then $M(\cdot, \omega)$ is continuous on $[0, \varphi(\omega))$, and continuous from the left at $\varphi(\omega)$.

(116) Digression. For simplicity, suppose $a = 0$ and $b = \infty$ for a moment. Confine ω to Ω_0. I say $\varphi \geqq \tau_0$. Indeed, suppose $\tau_0(\omega) > t$. Find α and β, depending on ω, with

$$0 < \alpha \leqq X(s, \omega) \leqq \beta < \infty \quad \text{for } 0 \leqq s \leqq t.$$

Remember that m is finite on $[\alpha, \beta]$, and use (96f):

$$\begin{aligned} M(t, \omega) &= \frac{1}{2}\int_{-\infty}^{\infty} \lambda(y, t, \omega)\, m(dy) \\ &= \frac{1}{2}\int_{\alpha}^{\beta} \lambda(y, t, \omega)\, m(dy) \\ &< \infty. \end{aligned}$$

If $m(0, \varepsilon) < \infty$, you can argue $\varphi = \infty$ by the same method. If $m(0, \varepsilon) = \infty$, theorems (132–134) show:

$$\int_0^{\varepsilon} x\, m(dx) < \infty \quad \text{implies} \quad B_x\{\varphi = \infty\} = 1 \text{ for } x > 0;$$

$$\int_0^{\varepsilon} x\, m(dx) = \infty \quad \text{implies} \quad B_x\{\varphi = \tau_0 \text{ and } M(\varphi) = \infty\} = 1 \text{ for } x > 0.$$

(117) Lemma. $B_x\{M(t) > 0\} = 1$ *for* $x \in K$ *and* $t > 0$.

PROOF. Fix $x \in K$. Let G be the set of $\omega \in G_0$ such that $\omega(0) = x$, and for all $t > 0$,

$$(118) \quad L(t, \omega) = \min\{\omega(s): 0 \leqq s \leqq t\} < \max\{\omega(s): 0 \leqq s \leqq t\} = U(t, \omega).$$

Then $B_x\{G\} = 1$ by (94) and (1.55). For $\omega \in G$, let $C(\omega)$ be the set of y such that $\lambda(y, dt, \omega)$ has support $S_y(\omega)$. Let H be the set of $\omega \in G$ such that m assigns measure 0 to the complement of $C(\omega)$. By (112) and Fubini: $C(\omega)$ and H are measurable; $B_x\{H\} = 1$.

Fix $\omega \in H$ and $t > 0$. I say $M(t, \omega) > 0$. Indeed, let $D(t, \omega)$ be the set of y such that

$$S_y(\omega) \cap [0, t) = \{u: 0 \leqq u < t \text{ and } \omega(u) = y\}$$

is nonempty. Check that $D(t, \omega)$ is an interval. Remember $\omega \in G$. Then

$$x \in K \cap (L(t, \omega), U(t, \omega)) \subset D(t, \omega),$$

using the assumption $x \in K$ and path continuity. This and (118) show: $D(t, \omega)$ includes a nonempty, relatively open subset of K; so $m\{D(t, \omega)\} > 0$. But m concentrates on $C(\omega)$, so

$$m\{C(\omega) \cap D(t, \omega)\} > 0.$$

From the definitions,

$$\lambda(y, t, \omega) > 0 \quad \text{for } y \in C(\omega) \cap D(t, \omega).$$

Then

$$M(t, \omega) \geqq \frac{1}{2} \int_{C(\omega) \cap D(t, \omega)} \lambda(y, t, \omega)\, m(dy) > 0. \qquad \bigstar$$

Let $M(\infty, \omega) = \lim_{t \to \infty} M(t, \omega)$. Let Ω_m be the set of $\omega \in \Omega_0$ such that: $M(\infty, \omega) = \infty$; while the intersection of the support of $M(dt, \omega)$ and $[0, \varphi(\omega))$ coincides with the intersection of $S_K(\omega)$ and $[0, \varphi(\omega))$.

(119) Theorem. $B_x\{\Omega_m\} = 1$ *for all* x.

PROOF. First, I say

$$B_x\{M(\infty) = \infty\} = 1.$$

Indeed, by (106c) and Fubini: for B_x-almost all ω,

$$m\{y: \lambda(y, \infty, \omega) < \infty\} = 0.$$

Second, if $\omega \in \Omega_0$, I say that the support of $M(dt, \omega)$ is included in $S_K(\omega)$. Indeed, suppose $(c, d) \cap S_K(\omega) = \emptyset$. Then $(c, d) \cap S_y(\omega) = \emptyset$ for all $y \in K$. So $\lambda(y, c, \omega) = \lambda(y, d, \omega)$ for all $y \in K$, by (111). And $M(c, \omega) = M(d, \omega)$, because m concentrates on K.

Third, let G be the set of $\omega \in \Omega_0$ such that

$$[0, \varphi(\omega)) \cap \text{support } M(dt, \omega) \supset [0, \varphi(\omega)) \cap S_K(\omega).$$

I say G is measurable and $B_x\{G\} = 1$. As in (112), let r be a nonnegative rational, and let $\tau(r)$ be the least $t \geqq r$ if any with $X(t) \in K$; let $\tau(r) = \infty$, if none. Let s be a positive rational. Then

$$G = \bigcap_{r,s} G(r, s),$$

where

$$G(r, s) = \{\Omega_0 \text{ and } \tau(r) \geqq \varphi\} \cup H(r, s)$$

and

$$H(r, s) = \{\Omega_0 \text{ and } \tau(r) < \varphi \text{ and } M[\tau(r) + s] > M[\tau(r)]\}.$$

Fix r and s. Abbreviate $\sigma = \tau(r)$ and $Y = X[\tau(r)]$. Use (114):

$$H(r, s) = \Omega_0 \cap \{\sigma < \varphi\} \cap \{S_\sigma \in [M(s) > 0]\}.$$

Clearly, σ and φ are Markov times; so $\{\sigma < \varphi\} \in \mathscr{B}(\sigma+)$ by (11a); and $Y \in K$ on $\{\sigma < \varphi\}$. Remember $B_x\{\Omega_0\} = 1$ by (94). Use strong Markov and (117):

$$\begin{aligned} B_x\{H(r, s)\} &= \int_{\{\sigma < \varphi\}} B_Y\{M(s) > 0\}\, dB_x \\ &= B_x\{\sigma < \varphi\}. \end{aligned}$$

So, $B_x\{G(r, s)\} = B_x\{\sigma \geqq \varphi\} + B_x\{\sigma < \varphi\} = 1$. ★

(120) Definition. *$T(t, \omega)$ is the greatest s such that $M(s, \omega) = t$, whenever $t < M[\varphi(\omega), \omega]$. Temporarily, leave $T(t, \omega)$ undefined for $t \geqq M[\varphi(\omega), \omega]$.*

(121) Lemma. *$(t, \omega) \to T(t, \omega)$ is measurable. For $\omega \in \Omega_m$, the function $T(\cdot, \omega)$ is nondecreasing and continuous from the right and $T(t, \omega) \in S_K(\omega)$ for all t.*

PROOF. Easy. ★

9. TRANSFORMATION OF TIME

Recall that $\{B_x\}$ is standard Brownian motion. The coordinate process X on Ω is Brownian motion starting from x, relative to B_x. Here Ω is the set of all continuous functions from $[0, \infty)$ to $(-\infty, \infty)$.

The case $I = (-\infty, \infty)$

The first theorem of this section constructs the general diffusion in natural scale on $(-\infty, \infty)$ from $\{B_x\}$ by transformation of time. Let m be a locally finite and strictly positive measure on $(-\infty, \infty)$. If m has a continuous density $\dot{m}$, as you should assume on a first reading, then (92) and (93) can be used in place of the local time properties developed in Section 8. Define M as in (113):

$$(122) \qquad M(t) = \frac{1}{2}\int_0^t \dot{m}[X(s)]\,ds = \frac{1}{2}\int_{-\infty}^{\infty} \lambda(y, t)\,m(dy).$$

You should think about the middle expression in (122) only for continuous $\dot{m}$.

The set Ω_0 was introduced in (94). The set Ω_m was introduced before (119). Under present circumstances, Ω_m consists of all $\omega \in \Omega_0$ such that: $M(\infty, \omega) = \infty$; while $M(\cdot, \omega)$ is strictly increasing on $[0, \infty)$. Automatically, $M(0, \omega) = 0$ and $M(\cdot, \omega)$ is continuous on $[0, \infty)$ for $\omega \in \Omega_0$. The assertion of (119) was: $B_x\{\Omega_m\} = 1$ for all x. Until further notice, confine ω to Ω_m. It is helpful to remember (1.17). Let $T(\cdot, \omega)$ be the inverse of $M(\cdot, \omega)$, as in (120). Then $T(0, \omega) = 0$, while $0 < T(t, \omega) < \infty$ for $0 < t < \infty$, but $T(\infty, \omega) = \infty$. Moreover, $T(\cdot, \omega)$ is continuous and strictly increasing. Finally, $T(t, \cdot)$ is measurable. Let $Y = X \circ T$; that is,

$$Y(t, \omega) = X[T(t, \omega), \omega].$$

Clearly, $Y(t, \cdot)$ is measurable, $Y(\cdot, \omega)$ is continuous, and $Y(0, \omega) = X(0, \omega)$. Let Q_x be the B_x-distribution of Y.

(123) Theorem. *$\{Q_x\}$ is a regular diffusion in natural scale on $(-\infty, \infty)$, with speed measure m.*

NOTE. By (123) and (66), a measure m on $(-\infty, \infty)$ is the speed measure of a regular diffusion $\{P_x\}$ in natural scale on $(-\infty, \infty)$ iff it is locally finite and strictly positive. Then $P_x = Q_x$ for all x by (90).

PROOF. There are only two interesting things to check, the strong Markov property and the speed measure.

Strong Markov property. Let $\mathscr{A}(t)$ be the σ-field spanned by $\{Y(s): s \leqq t\}$. Let τ be a random variable on Ω_m with values in $[0, \infty]$. Say τ is *Markov for* Y iff $\{\tau < t\} \in \mathscr{A}(t)$ for all $t \geqq 0$. Let $\mathscr{A}(\tau+)$ be the σ-field of measurable A such that $A \cap \{\tau < t\} \in \mathscr{A}(t)$ for all $t \geqq 0$.

As usual, $B_{Y(\tau)}Y^{-1}$ is the mapping

$$(\omega, A) \to B_{Y[\tau(\omega), \omega]}\{Y^{-1}A\}$$

of pairs (ω, A) for $\omega \in \{\tau < \infty\}$ and $A \in \mathscr{B}$. The thing to check is

(124) If τ is Markov for Y, then $B_{Y(\tau)}Y^{-1}$ is a regular conditional B_x-distribution for $Y(\tau + \cdot)$ given $\mathscr{A}(\tau+)$ on $\{\tau < \infty\}$.

I will do this in stages. Recall that $S_t\omega = \omega(t + \cdot)$, and let $\sigma(\omega) = T[\tau(\omega), \omega]$. Confine ω to $\Omega_m \cap \{\tau < \infty\}$ for (125–126). I claim

$$T[\tau(\omega) + t, \omega] = \sigma(\omega) + T[t, S_{\sigma(\omega)}\omega]. \tag{125}$$

As in Figure 21, let

$$\sigma(\omega) + \xi = T[\tau(\omega) + t, \omega], \quad \text{with } \xi = \xi(\omega).$$

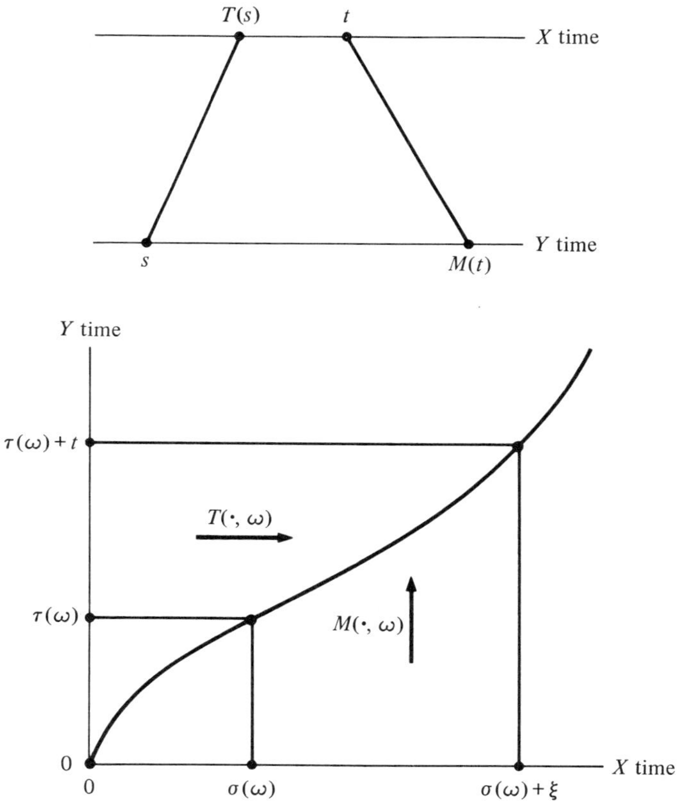

Figure 21.

Using (114),

$$\begin{aligned} \tau(\omega) + t &= M[\sigma(\omega) + \xi, \omega] \\ &= M[\sigma(\omega), \omega] + M[\xi, S_{\sigma(\omega)}\omega] \\ &= \tau(\omega) + M[\xi, S_{\sigma(\omega)}\omega]. \end{aligned}$$

So, $t = M[\xi, S_{\sigma(\omega)}\omega]$ and $\xi = T[t, S_{\sigma(\omega)}\omega]$, proving (125).

Next, I insist

(126) $$Y[\tau(\omega) + t, \omega] = Y[t, S_{\sigma(\omega)}\omega].$$

Indeed,

$$\begin{aligned} Y[\tau(\omega) + t, \omega] &= X\{T[\tau(\omega) + t, \omega], \omega\} \\ &= X\{\sigma(\omega) + T[t, S_{\sigma(\omega)}\omega], \omega\} \qquad \text{by (125)} \\ &= X\{T[t, S_{\sigma(\omega)}\omega], S_{\sigma(\omega)}\omega\} \\ &= Y[t, S_{\sigma(\omega)}\omega], \end{aligned}$$

proving (126).

Next, I maintain

(127) $$\{M(t) > r\} \cap \mathscr{A}(r) \subset \mathscr{B}(t).$$

For conventional reasons, I only have to get

$$\{M(t) > r\} \cap \{Y(s) < y\} \in \mathscr{B}(t)$$

for $0 \leqq s \leqq r$. Let $0 \leqq a \leqq b \leqq t$. Then

(128) $$\{a \leqq T(s) < b\} = \{M(a) \leqq s < M(b)\} \in \mathscr{B}(t),$$

because $M(u)$ is $\mathscr{B}(u)$-measurable. Let

$$T_n(s) = j/2^n \quad \text{when} \quad (j-1)/2^n \leqq T(s) < j/2^n.$$

If $j/2^n \leqq t$, then $\{T_n(s) = j/2^n\} \in \mathscr{B}(t)$ by (128). So

(129) $$\{T_n(s) < t \text{ and } X[T_n(s)] < y\} \in \mathscr{B}(t).$$

Clearly, $T_n(s) \downarrow T(s)$ as $n \uparrow \infty$. So,

$$\begin{aligned} \{M(t) > r\} \cap \{Y(s) < y\} &= \{T(r) < t\} \cap \{X[T(s)] < y\} \\ &= \textstyle\bigcup_{n=1}^{\infty} A_n, \end{aligned}$$

where

$$A_n = \{T_n(r) < t \text{ and } X[T(s)] < y\}.$$

Moreover,

$$A_n = \textstyle\bigcup_{j=n}^{\infty} \bigcap_{i=j}^{\infty} C_{ni},$$

where

$$C_{ni} = \{T_n(r) < t \text{ and } X[T_i(s)] < y\}.$$

Remember $s \leqq r$. If $i \geqq n$ and $T_n(r) < t$, then $T_i(s) \leqq T_n(s) \leqq T_n(r) < t$. So

$$C_{ni} = \{T_n(r) < t\} \cap \{T_i(s) < t \text{ and } X[T_i(s)] < y\}$$

is in $\mathscr{B}(t)$ by (129). This completes the proof of (127).

Next, I assert

(130) $\qquad\sigma$ is a Markov time.

Indeed, $\sigma < t$ iff $\tau = M(\sigma) < M(t)$. But

$$\{\tau < M(t)\} = \bigcup_{\text{rational } r} \{\tau < r < M(t)\},$$

and $\{\tau < r\} \in \mathscr{A}(r)$ because τ is Markov for Y. Use (127) to complete the proof of (130).

Next, I say

(131) $$\mathscr{A}(\tau+) \subset \mathscr{B}(\sigma+).$$

Indeed, let $A \in \mathscr{A}(\tau+)$. Then

$$A \cap \{\sigma < t\} = A \cap \{\tau < M(t)\}$$
$$= \bigcup_{\text{rational } r} A \cap \{\tau < r < M(t)\}.$$

But $A \cap \{\tau < r\} \in \mathscr{A}(r)$, because $A \in \mathscr{A}(\tau+)$. Use (127) to complete the proof of (131). Presumably, equality holds in (131).

I can now argue (124). Let $C \in \mathscr{B}$. By (126),

$$\{Y(\tau + \cdot) \in C\} = \{S_\sigma \in [Y \in C]\}.$$

Let $A \in \mathscr{A}(\tau+)$ and $A \subset \{\tau < \infty\}$. Then $A \in \mathscr{B}(\sigma+)$ by (131). By thinking, $A \subset \{\sigma < \infty\}$. Clearly, $X(\sigma) = Y(\tau)$. Use strong Markov on $\{B_x\}$.

$$B_x\{A \text{ and } Y(\tau + \cdot) \in C\} = B_x\{A \text{ and } S_\sigma \in [Y \in C]\}$$
$$= \int_A B_{X(\sigma)}[Y \in C]\, dB_x$$
$$= \int_A B_{Y(\tau)}[Y \in C]\, dB_x.$$

You should now check that $\{Q_x\}$ is a regular diffusion in natural scale on $(-\infty, \infty)$; lemma (1.114) will help. Finally, I will verify that the

Speed measure is m. Suppose $a < b$ and $a \leqq x \leqq b$. Let $J = [a, b]$. Recall that τ_J is the least t if any with $X(t) = a$ or b, and $\tau_J = \infty$ if none. Then $M(\tau_J)$ is the least t if any with $Y(t) = a$ or b, and $M(\tau_J) = \infty$ if none. Check this computation.

$$\begin{aligned}
E_x\{M(\tau_J)\} &= E_x\left\{\frac{1}{2}\int_{-\infty}^{\infty} \lambda(y,\tau_J)\, m(dy)\right\} && \text{by (122)}\\
&= E_x\left\{\frac{1}{2}\int_J \lambda(y,\tau_J)\, m(dy)\right\} && \text{by (96f)}\\
&= \frac{1}{2}\int_J E_x\{\lambda(y,\tau_J)\}\, m(dy) && \text{by Fubini}\\
&= \int_J G_J(x,y)\, m(dy) && \text{by (97), or (92) in the good case.}
\end{aligned}$$

Now use (66); the differentiation was done in (58). ★

The case $I = (0, \infty)$

The next theorem constructs the general diffusion in natural scale on $(0, \infty)$ from $\{B_x\}$ by transformation of time. Let m be a locally finite and strictly positive measure on $(0, \infty)$, with $\int_0^\varepsilon x\, m(dx) = \infty$ for $\varepsilon > 0$. Let $\tau = \tau_0$ be the least t if any with $X(t) = 0$, and $\tau = \infty$ if none. Let

$$M(t) = \frac{1}{2}\int_{-\infty}^{\infty} \lambda(y,t)\, m(dy) \quad \text{for } t \leqq \tau$$

$$M(t) = M(\tau) \qquad \text{for } t > \tau.$$

Suppose $\omega \in \Omega_0$. Then $M(0, \omega) = 0$. Moreover, $M(\cdot, \omega)$ is finite, continuous, and nondecreasing on $[0, \tau(\omega))$, and even continuous from the left at $\tau(\omega)$. The possibility $M[\tau(\omega), \omega] = \infty$ is strong. Let Ω_M be the set of $\omega \in \Omega_0$ such that $M(\cdot, \omega)$ is strictly increasing on $[0, \tau(\omega))$. As (119) implies, $B_x(\Omega_M) = 1$ for all $x \geqq 0$. Confine ω to $\{\Omega_M \text{ and } \tau < \infty\}$ until further notice. Let

$$T(t, \omega) = M(\cdot, \omega)^{-1}(t) \quad \text{for } t < M[\tau(\omega), \omega]$$

$$T(t, \omega) = \tau(\omega) \qquad \text{for } t \geqq M[\tau(\omega), \omega].$$

As before, $T(0, \omega) = 0$, and $0 < T(t, \omega) < \infty$ for $0 < t < M[\tau(\omega), \omega]$. Further, $T(\cdot, \omega)$ is continuous, and strictly increasing on $[0, M(\tau(\omega), \omega)]$. Finally, $T(t, \cdot)$ is measurable. Let $Y = X \circ T$; that is,

$$Y(t, \omega) = X[T(t, \omega), \omega].$$

Clearly, $Y(t, \cdot)$ is measurable, $Y(\cdot, \omega)$ is continuous, and $Y(0, \omega) = X(0, \omega)$. Let Q_x be the B_x-distribution of Y, for $x \geqq 0$.

(132) Theorem. $B_x\{M(\tau) = \infty\} = 1$ *for* $x > 0$. *Moreover,* $\{Q_x : x > 0\}$ *is a regular diffusion in natural scale on* $(0, \infty)$, *with speed measure* m.

NOTE. By (132), (66), and (78), a measure m on $(0, \infty)$ is the speed measure of a regular diffusion $\{P_x\}$ in natural scale on $(0, \infty)$ iff m is locally finite, strictly positive, and $\int_0^\varepsilon x\, m(dx) = \infty$ for $\varepsilon > 0$. Then $P_x = Q_x$ for all $x > 0$ by (90).

PROOF. The argument for (123) shows that $\{Q_x : x \geqq 0\}$ is a diffusion on $[0, \infty)$, with 0 absorbing. It may not be regular. For $0 < a < x < b$,

$$Q_x\{\text{hit } b \text{ before } a\} = \frac{x - a}{b - a} \quad \text{and} \quad Q_x\{\text{hit } a \text{ before } b\} = \frac{b - x}{b - a}.$$

I claim

(133) $Q_x\{\text{hit } 0\}$ is 0 or 1 and does not depend on $x > 0$.

Indeed, let $0 < x < y$. Then

$$Q_y\{\text{hit } 0\} = Q_y\{\text{hit } x\} \cdot Q_x\{\text{hit } 0\}$$

by (10). But $Q_y\{\text{hit } x\} = 1$, as in (76). This proves that $Q_x\{\text{hit } 0\}$ is a constant c for $0 < x < \infty$. If $c > 0$, then $\{Q_x : x \geqq 0\}$ is regular on $[0, \infty)$. For $0 < x \leqq 1$,

$$c = Q_x\{\text{hit } 0\} \geqq Q_x\{\text{hit } 0 \text{ before } 1\} \to 1$$

as $x \to 0$ by (26); so $c = 1$. This proves (133).

If $Q_x\{\text{hit } 0\} = 1$, then $\{Q_x : x \geqq 0\}$ is regular on $[0, \infty)$. The speed measure is m on $(0, \infty)$, as in (123). Therefore $\int_0^\varepsilon x\, m(dx) < \infty$, by (67a). This contradiction shows $Q_x\{\text{hit } 0\} = 0$, that is, $B_x\{M(\tau) = \infty\} = 1$, for all $x > 0$. Thus, $\{Q_x : x > 0\}$ is a regular diffusion in natural scale on $(0, \infty)$. As in (123), the speed measure is m. ★

The case $I = [0, \infty)$ and 0 is absorbing

The next theorem constructs the general diffusion in natural scale on $[0, \infty)$, with 0 absorbing, from $\{B_x\}$ by transformation of time. Let m be a locally finite and strictly positive measure on $(0, \infty)$, with $\int_0^\varepsilon x\, m(dx) < \infty$ for $\varepsilon > 0$. Define τ, M, T, Y and Q as for (132).

(134) Theorem. $B_x\{M(\tau) < \infty\} = 1$ *for* $x \geqq 0$. *Moreover,* $\{Q_x : x \geqq 0\}$ *is a regular diffusion in natural scale on* $[0, \infty)$, *with* 0 *absorbing, and speed measure* m.

NOTE. By (134), (66), and (67a), a measure m on $(0, \infty)$ is the speed measure of a regular diffusion $\{P_x\}$ in natural scale on $[0, \infty)$, with 0 absorbing, iff m is locally finite, strictly positive, and $\int_0^\varepsilon x\, m(dx) < \infty$ for $\varepsilon > 0$. Then $P_x = Q_x$ for all $x \geqq 0$ by (90).

PROOF. As in (132), using (78) instead of (67a). ★

The case $I = [0, \infty)$ and 0 is instantaneous

The next theorem constructs the general diffusion in natural scale on $[0, \infty)$, with 0 instantaneous, from $\{B_x\}$ by transformation of time. Let m be a locally finite and strictly positive measure on $[0, \infty)$. In particular, $m[0, \varepsilon] < \infty$ for $\varepsilon < 0$. As in (113), let

$$M(t) = \frac{1}{2}\int_{[0,\infty)} \lambda(y, t)\, m(dy).$$

As (119) implies, $B_x\{\Omega_m\} = 1$ for all x. Confine ω to Ω_m for the discussion of (135). Then $M(0, \omega) = 0$, and $0 < M(t, \omega) < \infty$ for $0 < t < \infty$, while $M(\infty, \omega) = \infty$. Moreover, $M(\cdot, \omega)$ is continuous, nondecreasing, and has support $S_{[0,\infty)}(\omega)$. As in (120), let $T(t, \omega) = \max\{s : M(s, \omega) = t\}$. Now (121) implies $T(0, \omega) = 0$, and $0 < T(t, \omega) < \infty$ for $0 < t < \infty$, while $T(\infty, \omega) = \infty$. Moreover, $T(\cdot, \omega)$ is continuous from the right and nondecreasing. Finally, $T(t, \cdot)$ is measurable and $T(t, \omega) \in S_{[0,\infty)}(\omega)$. Let $Y = X \circ T$, that is,

$$Y(t, \omega) = X[T(t, \omega), \omega].$$

Check that $Y(t, \cdot)$ is measurable, $Y(\cdot, \omega)$ is continuous, $Y(t, \omega) \geqq 0$ and $Y(0, \omega) = X(0, \omega)$. For $x \geqq 0$, let Q_x be the B_x-distribution of Y.

(135) Theorem. *$\{Q_x : x \geqq 0\}$ is a regular diffusion in natural scale on $[0, \infty)$, with 0 instantaneous, and speed measure m.*

NOTE. By (135), (66), and (82), a measure m on $[0, \infty)$ is the speed measure of a regular diffusion $\{P_x\}$ in natural scale on $[0, \infty)$ with 0 instantaneous, iff m is locally finite and strictly positive. Then $P_x = Q_x$ for all $x \geqq 0$ by (90).

PROOF. Use the argument for (123), with easy modifications, to check that $\{Q_x : x \geqq 0\}$ is a regular diffusion in natural scale on $[0, \infty)$, with speed measure m on $(0, \infty)$. The speed measure of 0 was defined in (88). I will now show that the speed measure of 0 is $m(0)$.

Let $\varepsilon > 0$. Recall that τ_ε is the least t if any with $X(t) = \varepsilon$, and $\tau_\varepsilon = \infty$ if none. Recall that $\tau_{[-N,\varepsilon]}$ is the least t if any with $X(t) = -N$ or ε, and $\tau_{[-N,\varepsilon]} = \infty$ if none. Remember G_J from (57). Use Fubini, (96f), and (97 or 92):

$$\begin{aligned} E_0\{M(\tau_{[-N,\varepsilon]})\} &= E_0\left\{\frac{1}{2}\int_{-\infty}^{\infty} \lambda(y, \tau_{[-N,\varepsilon]})\, m(dy)\right\} \\ &= \int_{[0,\varepsilon]} G_{[-N,\varepsilon]}(0, y)\, m(dy) \\ &= \frac{N}{N + \varepsilon}\int_{[0,\varepsilon]} (\varepsilon - y)\, m(dy). \end{aligned}$$

Let $N \uparrow \infty$. Then $\tau_{[-N,\varepsilon]} \uparrow \tau_\varepsilon$, so $M(\tau_{[-N,\varepsilon]}) \uparrow M(\tau_\varepsilon)$. By monotone convergence and Fubini,

$$\begin{aligned}\int \tau_\varepsilon \, dQ_0 &= E_0\{M(\tau_\varepsilon)\} \\ &= \lim_{N\to\infty} E_0\{M(\tau_{[-N,\varepsilon]})\} \\ &= \int_{[0,\varepsilon]} (\varepsilon - y)\, m(dy) \\ &= \int_{0 \leqq y \leqq \varepsilon} \int_{y \leqq z \leqq \varepsilon} dz\, m(dy) \\ &= \int_{[0,\varepsilon]} m[0, z]\, dz.\end{aligned}$$

The speed measure of 0 is

$$\lim_{\varepsilon\to 0} \frac{1}{\varepsilon} \int \tau_\varepsilon \, dQ_0 = m(0). \qquad \bigstar$$

The other cases

Any halfline can be reduced to $(0, \infty)$ or $[0, \infty)$ by appropriate change of scale. Regular diffusions in natural scale on halflines are therefore covered by this section. The results, appropriately modified, also apply to finite intervals.

Notice

The remaining sections of this chapter cover more specialized topics, and are quite lean.

10. DIFFUSION LOCAL TIME

Continue with the notation of Section 9; for simplicity, assume I is unbounded and there are no absorbing points. So, I is one of $(-\infty, \infty)$, $(0, \infty)$, $[0, \infty)$. And m is a strictly positive, locally finite measure on I. If $I = (0, \infty)$, assume $\int_0^\varepsilon x\, m(dx) = \infty$. If $I = (-\infty, \infty)$ or $[0, \infty)$,

$$M(t) = \frac{1}{2} \int_{-\infty}^{\infty} \lambda(t, y)\, m(dy)$$

$$T = M^{-1}, \quad \text{in the sense of (120).}$$

Remember that τ_0 is the time to hit 0. If $I = (0, \infty)$,

$$M(t) = \frac{1}{2}\int_{-\infty}^{\infty} \lambda(t, y)\, m(dy) \quad \text{for } t \leqq \tau_0$$

$$B_x\{M(\tau_0) = \infty\} = 1 \quad \text{for } x > 0$$

$$T = M^{-1}, \quad \text{in the sense of (120).}$$

If $I = (-\infty, \infty)$ or $[0, \infty)$, then Ω_m is the set of $\omega \in \Omega_0$ such that $M(\infty, \omega) = \infty$ and the support of $M(dt, \omega)$ is $S_I(\omega)$; and $B_x\{\Omega_m\} = 1$ for all x by (119); confine ω to Ω_m. If $I = (0, \infty)$, let Ω_m be the set of $\omega \in \Omega_0$ such that $\omega(0) > 0$ and $\tau_0(\omega) < \infty$ and $M[\tau_0(\omega), \omega] = \infty$ and $M(\cdot, \omega)$ is strictly increasing on $[0, \tau_0(\omega))$; and $B_x\{\Omega_m\} = 1$ for all $x > 0$ by (119, 132); confine ω to Ω_m.

Let $Y = X \circ T$; that is,

$$Y(t, \omega) = X[T(t, \omega), \omega].$$

Let Q_x be the B_x-distribution of Y, for $x \in I$. As (123, 132, 135) prove, Q is a regular diffusion in natural scale on I, with speed measure m.

Let

$$\lambda^*(y, t, \omega) = \lambda[y, T(t, \omega), \omega].$$

(136) Theorem. *The set of ω where* (137) *holds is measurable and has B_x-measure* 1, *for $x \in I$.*

(137) *For all Borel $A \subset I$ and all $t \geqq 0$,*

$$\text{Lebesgue } \{s : 0 \leqq s \leqq t \text{ and } Y(s, \omega) \in A\} = \frac{1}{2}\int_A \lambda^*(y, t, \omega)\, m(dy).$$

PROOF. To begin with, A can be confined to rational intervals, and t to rationals, so (137) describes a measurable set of ω. I only have to check that for $\omega \in \Omega_m$ and $t \geqq 0$ and $f \in BC(I)$,

$$\int_0^t f[Y(s, \omega)]\, ds = \frac{1}{2}\int_I f(y)\lambda^*(y, t, \omega)\, m(dy). \tag{138}$$

Here is a computation.

$$\begin{aligned}\int_0^t f\{X[T(s, \omega), \omega]\}\, ds &= \int_0^{T(t,\omega)} f[X(u, \omega)]\, M(du, \omega)\\ &= \frac{1}{2}\int_{-\infty}^{\infty}\int_0^{T(t,\omega)} f[X(u, \omega)]\lambda(y, du, \omega)\, m(dy)\\ &= \frac{1}{2}\int_{-\infty}^{\infty} f(y)\lambda[y, T(t, \omega), \omega]\, m(dy).\end{aligned}$$

To check line 1, put $u = T(s, \omega)$; argue that $T(\cdot, \omega)$ carries Lebesgue measure on $[0, t]$ into $M(du, \omega)$ on $[0, T(t, \omega)]$, because $T = M^{-1}$. To check line 2, argue as in (1.112):

$$\int_0^\theta g(u)\, M(du, \omega) = \frac{1}{2}\int_{-\infty}^{\infty}\int_0^\theta g(u)\lambda(y, du, \omega)\, m(dy)$$

is obvious when g is the indicator of $[0, c]$, and extends to $g \in BC[0, \infty)$ by linearity and monotone convergence. Line 3 follows from (111): for $\omega \in \Omega_0$, the measure $\lambda(y, du, \omega)$ concentrates on $S_y(\omega) = \{u : X(u, \omega) = y\}$. ★

As a corollary, I claim:

$Q_x\{\text{Lebesgue } S_x > 0\}$ is one or zero, according as $m\{x\}$ is positive or zero.

Indeed,

$$\begin{aligned} Q_x\{\text{Lebesgue } S_x > 0\} &= B_x[\text{Lebesgue } \{t : Y(t) = x\} > 0] \\ &= B_x[m\{x\} \cdot \lambda^*(x, \infty) > 0] \qquad \text{by (137)} \\ &= B_x[m\{x\} \cdot \lambda(x, T(\infty, \cdot)) > 0] \qquad \text{by definition.} \end{aligned}$$

Check $T(\infty, \cdot) > 0$, and use (106b) to complete the argument. This explains the difference between slow and fast reflection, as defined in (88). ★

By (1.58), for any x,

(139) For Q_x-almost ω, the set $S_x(\omega)$ is perfect and nowhere dense.

According to (Itô and McKean, 1965, p. 149 and 175),

(140) Theorem. *The Q_x-distribution of $X(t)$ has a density $y \to p(t, x, y)$ with respect to m, such that $(t, x, y) \to p(t, x, y)$ is continuous, $p(t, x, y) = p(t, y, x) > 0$,*

$$\frac{\partial}{\partial t} p(t, x, y) = [\Gamma p(t, x, \cdot)](y),$$

and

$$\frac{1}{2}\frac{\partial}{\partial t} E_x[\lambda^*(t, y)] = p(t, x, y).$$

11. FIRST EXAMPLES

Standard Brownian motion was discussed in (91). Brownian motion on $[0, \infty)$ with a reflecting barrier at 0, as defined in (1.40), is a regular diffusion in natural scale on $[0, \infty)$ with speed measure equal to twice Lebesgue measure on $[0, \infty)$. Here is what theorem (135) says in this case. Take normalized Brownian motion starting from $x \geqq 0$, but skip over the times at which the process is negative. The result is distributed like the absolute value of normalized Brownian motion starting from x. Brownian motion on $[0, \infty)$ with an absorbing barrier at 0, as defined in (1.47a), is a regular diffusion in natural scale on $[0, \infty)$ with speed measure equal to twice Lebesgue measure on $(0, \infty)$, and equal to ∞ by convention at 0.

A more interesting example is Brownian motion with *drift*. Let a be a negative real number. Recall that $\{B_x\}$ is standard Brownian motion. Let A_x be the B_x-distribution of Y, where

$$Y(t) = X(t) + at$$

for $0 \leqq t < \infty$. Clearly, $\{A_x\}$ is a regular diffusion on $(-\infty, \infty)$, but is not in natural scale. Following De Moivre, I will find a real number k so that

$$Z(t) = e^{kY(t)} = e^{kat}\, e^{kX(t)}$$

is a martingale with respect to each B_x. That is,

$$1 = E_0\{e^{kX(t)}\}\, e^{kat} = e^{\frac{1}{2}k^2t + kat},$$

or

$$0 = \tfrac{1}{2}k^2 t + kat$$

and

$$k = -2a > 0.$$

Arguing as in (1.114),

(141) $$\{A_x\} \text{ has scale function } s(x) = e^{-2ax}.$$

Of course, $s(I) = (0, \infty)$ and $s^{-1}x = -(1/2a)\log x$ for $0 < x < \infty$. For $0 < x < \infty$, let C_x be the $A_{s^{-1}x}$-distribution of $s(X)$. More explicitly, C_x is the $B_{s^{-1}x}$-distribution of Z, where

$$Z(t) = s[Y(t)] = e^{-2a^2t}\, e^{-2aX(t)}$$

for $0 \leqq t < \infty$. Then $\{C_x : 0 < x < \infty\}$ is a regular diffusion in natural scale on $(0, \infty)$. Let $0 < \alpha < \beta < \infty$, and let σ be the least t with $Z(t) = \alpha$ or β. Let $\alpha \leqq x \leqq \beta$. The next problem is to compute $E_{s^{-1}x}(\sigma)$. Now

$$Y(\sigma) = X(\sigma) + a\sigma;$$

so

$$E_{s^{-1}x}(\sigma) = \frac{1}{a} E_{s^{-1}x}\{Y(\sigma)\} - \frac{1}{a} E_{s^{-1}x}\{X(\sigma)\}.$$

Arguing as in (1.114),

$$E_{s^{-1}x}\{X(\sigma)\} = s^{-1}x = -\frac{1}{2a}\log x.$$

On the other hand,

$$Y(\sigma) = s^{-1}\alpha \quad \text{when } Z(\sigma) = \alpha$$

$$Y(\sigma) = s^{-1}\beta \quad \text{when } Z(\sigma) = \beta.$$

Because Z is in natural scale,

$$E_{s^{-1}x}[Y(\sigma)] = -\frac{x-\alpha}{\beta-\alpha}\frac{1}{2a}\log\beta - \frac{\beta-x}{\beta-a}\frac{1}{2a}\log\alpha.$$

Differentiate twice to get

(142) $\{C_x\}$ has speed measure m with $m(dx) = \dfrac{dx}{2a^2x^2}$.

12. AN EXAMPLE OF FELLER AND McKEAN

This section presents the first example (Feller and McKean, 1956) of a countable state Markov chain with standard transitions and all states instantaneous. Namely, let $D = \{d_1, d_2, \ldots\}$ be a countable dense subset of $(-\infty, \infty)$. Let the measure m on $(-\infty, \infty)$ concentrate on D and assign positive mass to d_n, with $\Sigma\, m(d_n) < \infty$. Clearly, m is strictly positive and locally finite. As in Section 9, construct from standard Brownian motion $\{B_x\}$ the diffusion $\{Q_x\}$ in natural scale on $(-\infty, \infty)$, with speed measure m. Briefly, Ω is the set of functions from $[0, \infty)$ to $(-\infty, \infty)$, and X is the coordinate process on Ω. Define M by (113). Let $T = M^{-1}$ as in (120), and $Y = X \circ T$. That is, $T(\cdot, \omega)$ is the inverse function of $M(\cdot, \omega)$, and $Y(t, \omega) = X[T(t, \omega), \omega]$. Let Q_x be the B_x-distribution of Y. As (123) implies, $\{Q_x\}$ is a regular diffusion in natural scale on $(-\infty, \infty)$, with speed measure m. For a and b in D and $t \geqq 0$, let

$$Q(t, a, b) = B_a\{Y(t) = b\}.$$

(143) Theorem. *Q is a standard stochastic semigroup on D, with all states instantaneous.*

This would be immediate from (139) and (140). I didn't prove general (140), so I have to do the relevant special case.

(144) Lemma. *For $a \in D$ and $t \geqq 0$,*

(145) $$B_a\{Y(t) \in D\} = 1.$$

PROOF. Let G_a be the set of $t \geqq 0$ where (145) holds. Clearly, $0 \in G_a$. By (136) and Fubini, G_a is Borel and its complement has Lebesgue measure 0. Let $G = \bigcap_{a \in D} G_a$. Then G is Borel, $0 \in G$, the complement of G has zero

Lebesgue measure, and G is a semigroup by the Markov property: more explicitly, suppose s and t are in G. Then

$$\begin{aligned} B_a\{Y(s+t)\in D\} &= B_a\{Y(s)\in D \text{ and } Y(s+t)\in D\} \\ &= \Sigma_{d\in D}\, B_a\{Y(s)=d \text{ and } Y(s+t)\in D\} \\ &= \Sigma_{d\in D}\, B_a\{Y(s)=d\}\cdot B_d\{Y(t)\in D\} \\ &= 1, \end{aligned}$$

where line 3 is justified by strong Markov on Brown, and (126) with $\tau \equiv s$. Consequently, $G = [0, \infty)$ by (3.15) of *ACM*. ★

(146) Corollary. *Q is a stochastic semigroup on D.*

PROOF. Use (123), (144), and lemma (5.4) of *MC*. ★

Temporarily, give D the discrete topology, and compactify by adjoining φ. Let

$$\begin{aligned} Y^*(t,\omega) &= Y(t,\omega) \quad \text{for } Y(t,\omega)\in D \\ &= \varphi \qquad\qquad \text{for } Y(t,\omega)\notin D. \end{aligned}$$

Remember Ω_m from (119). Let Ω^* be the set of $\omega\in\Omega_m$ such that: $Y(t,\omega)\in D$ for all binary rational t; and the measure $\lambda(a, dt, \omega)$ has support $S_a(\omega) = \{t:\omega(t)=a\}$ for all $a\in D$; and $S_a(\omega)$ is Lebesgue null for all $a\in D$.

NOTE. $\lambda(a, dt, \omega)$ is continuous, and therefore has perfect support. If $S_a(\omega)$ is Lebesgue null, it includes no proper interval.

Use (119, 144, 112) and (1.58) to see $B_a\{\Omega^*\} = 1$ for all $a\in D$.

(147) Lemma. *For $\omega\in\Omega^*$, the function $Y^*(\cdot,\omega)$ is quasiregular in the sense of (MC, 9.7), and constant on no interval.*

PROOF. Easy. ★

(148) Lemma. *Q is standard.*

PROOF. Suppose $d\in D$ and $\omega\in\Omega^*$ and $\omega(0)=d$. As (136) implies,

$$\text{Lebesgue } \{s: 0 \leqq s \leqq t \text{ and } Y(s,\omega)=d\} = \tfrac{1}{2}m(d)\lambda^*(y,t,\omega) > 0.$$

Now use (1.17) of *ACM*.

Use (*MC*, 9.18) to complete the proof of (143). Parenthetically, this process has no pseudo-jumps by continuity. Its generator must vanish off the diagonal by (2.1) of *ACM*.

13. AN EXAMPLE OF BREIMAN

The process described in this section was suggested by Leo Breiman. Let

$$d(1) > d(2) > \cdots > 0 \quad \text{and} \quad d(n) \to 0.$$

Let $D = \{d(1), d(2), \ldots\}$ and $C = D \cup \{0\}$. So C is closed. Let m be a measure on C, with $m\{0\} = 0$ and $m\{d(n)\} > 0$ and $\Sigma_{n=1}^{\infty} m\{d(n)\} < \infty$. The support of m is C. Remember that Ω is the set of continuous functions from $[0, \infty)$ to $(-\infty, \infty)$; the probability B_x on Ω makes the coordinate process X Brownian motion starting from x. Remember

$$S_x(\omega) = \{t : \omega(t) = x\} \quad \text{and} \quad S_A(\omega) = \{t : \omega(t) \in A\}.$$

Let Ω^* be the set of $\omega \in \Omega_0$, as defined in (94), such that

$$\limsup_{t \to \infty} \omega(t) = \infty \quad \text{and} \quad \liminf_{t \to \infty} \omega(t) = -\infty,$$

each point in $S_0(\omega)$ is the limit both from the right and left of sequences in $S_{(0,\infty)}(\omega)$,

and for all $c \in C$:

the measure $\lambda(c, dt, \omega)$ has support $S_c(\omega)$;

$S_c(\omega)$ has Lebesgue measure 0;

$\lambda(c, \infty, \omega) = \infty$.

Using (94), (1.17), (1.57e, f), (112), (1.58), and (106), verify $B_c\{\Omega^*\} = 1$ for all $c \in C$. Confine ω to Ω^* for the rest of this section. Define M by (113):

$$M(t) = \tfrac{1}{2} \Sigma_{c \in C}\, \lambda(c, t) m\{c\}.$$

Define T by (120), as the right continuous inverse of M. Let $Y = X \circ T$, namely,

$$Y(t, \omega) = X\{T(t, \omega), \omega\}.$$

Let

$$P(t, a, b) = B_a\{Y(t) = b\}$$

for $a, b \in D$. Clearly, D is discrete; and C is the one-point compactification of D. With this understanding,

(149) Theorem. *P is a standard stochastic semigroup on D, with all states stable. Let $d \in D$. With respect to B_d, the process Y is Markov with transitions P and starting state d. The Y-sample functions are regular in the sense of (MC, 7.2), with 0 as the fictitious state φ.*

The process Y commits various atrocities. Before describing them, or proving (149), I want to look more closely at the Y-sample functions. Remember that ω is confined to Ω^*. Then $M(0, \omega) = 0$ and $0 < M(t, \omega) < \infty$ for $0 < t < \infty$ and $M(\infty, \omega) = \infty$. Consequently, $T(0, \omega) = 0$ and $0 < T(t, \omega) < \infty$ for $0 < t < \infty$ and $T(\infty, \omega) = \infty$.

Check

(150) If $d \in D$, then $S_d(\omega)$ is isolated from the rest of $S_C(\omega)$.

(151) If $t \in S_0(\omega)$ and $\varepsilon > 0$, then $(t - \varepsilon, t)$ and $(t, t + \varepsilon)$ both meet $S_d(\omega)$ for arbitrarily small $d \in D$.

If K is a closed set, let K^* be K deprived of its right endpoints: x is a *right endpoint* of K iff $x \in K$ but $(x, x + \varepsilon)$ is free of K for small positive ε; alternatively, x is a right endpoint of K iff x is the left endpoint of a maximal interval complementary to K—then write $\rho(x, K)$ for the right endpoint of this maximal interval.

Remember that the measure $M(dt, \omega)$ has support $S_C(\omega)$; so

(152) $M(\cdot, \omega)$ is a continuous, strictly increasing map of $S_C(\omega)^*$ onto the time scale $[0, \infty)$ of $Y(\cdot, \omega)$. If $t \in S_C(\omega)^*$, then

$$T[M(t, \omega), \omega] = t$$

and

$$Y[M(t, \omega), \omega] = X(t, \omega).$$

(153) If t is a right endpoint of $S_C(\omega)$, then $T[M(t, \omega), \omega] = \rho(t, S_c(\omega))$. Necessarily $t \in S_d(\omega)$ for some $d \in D$; and $\rho(t, S_c(\omega))$ is the least $s > t$ with $X(s, \omega) \in D$. If $d = d(1)$, then $Y(t, \omega) = d(1)$ or $d(2)$. If $d = d(n)$ for $n > 1$, then $Y(t, \omega) = d(n + 1)$ or $d(n)$ or $d(n - 1)$.

Let

$$U_c(\omega) = \{t: Y(t, \omega) = c\}.$$

Because $\omega \in \Omega^*$,

(154) $S_c(\omega)$ is perfect and unbounded.

Use (151–152):

(155) $U_0(\omega)$ is the $M(\cdot, \omega)$-image of $S_0(\omega)$.

Use (154), and (1.30) of *ACM*.

(156) $U_0(\omega)$ is perfect, unbounded, and of Lebesgue measure 0.

Use (151–152) again:

(157) $Y(\cdot, \omega)$ is continuous at all $t \in U_0(\omega)$.

For each $d \in D$, the maximal complementary intervals of $S_d(\omega)$ which meet $S_c(\omega)$ do not accumulate, by (150). This and (152–153) force:

(158) On each maximal interval complementary to $U_0(\omega)$, the function $Y(\cdot, \omega)$ is a right continuous step function; the only jumps which occur are from $d(1)$ to $d(2)$, and from $d(n)$ to $d(n+1)$ or $d(n-1)$ when $n \geqq 2$.

PROOF OF (149). First, argue that $B_d\{Y(t) \in D\} = 1$: use (156) and the method of (144). Then argue that the B_c-distribution of $Y(t + \cdot)$ given $Y(s)$ for $0 \leqq s \leqq t$ and $U(t) = d$ is B_d: use the method for (123). This makes P a stochastic semigroup on D and the transitions for Y, by (MC, 5.4). Standardness and regularity come from (157–158). So does stability, using (MC, 9.18). ★

Abbreviate $\sigma_1 = \tau_{d(2)}$, the time for X to hit $d(2)$; and for $n \geqq 2$, let

$$\sigma_n = \tau_{[d(n+1), d(n-1)]},$$

the time for X to hit $d(n+1)$ or $d(n-1)$. I claim:

(159) If $X(0, \omega) = d(1)$, then $Y(u, \omega) = d(1)$ for

$$0 \leqq u < M[\sigma_1(\omega), \omega] = \tfrac{1}{2}m\{d(1)\}\lambda[d(1), \sigma_1(\omega), \omega];$$

and if $\sigma_1(\omega) < \infty$,

$$Y\{M[\sigma_1(\omega), \omega], \omega\} = X\{\sigma_1(\omega), \omega\} = d(2).$$

For $n \geqq 2$, I claim:

(160) If $X(0, \omega) = d(n)$, then $Y(u, \omega) = d(n)$ for

$$0 \leqq u < M[\sigma_n(\omega), \omega] = \tfrac{1}{2}m\{d(n)\}\lambda[d(n), \sigma_n(\omega), \omega];$$

and if $\sigma_n(\omega) < \infty$,

$$Y\{M[\sigma_n(\omega), \omega], \omega\} = X\{\sigma_n(\omega), \omega\} = d(n+1) \quad \text{or} \quad d(n-1).$$

I will only argue (160), because (159) is similar and easier. Suppose $X(0, \omega) = d(n)$ and $0 \leqq t < \sigma_n(\omega)$. Then

$$M(t, \omega) = \tfrac{1}{2}m\{d(n)\}\lambda[d(n), t, \omega]$$

by (96). But $\lambda[d(n), dt, \omega]$ has support $S_{d(n)}(\omega)$. So $T(u, \omega) \in S_{d(n)}(\omega)$ and $Y(u, \omega) = d(n)$ for $0 \leqq u < M[\sigma_n(\omega), \omega]$. Suppose $\sigma_n(\omega) < \infty$. Then $M(\cdot, \omega)$ is strictly increasing at $\sigma_n(\omega) \in S_{d(n+1)}(\omega) \cup S_{d(n-1)}(\omega)$, and

$$T\{M[\sigma_n(\omega), \omega], \omega\} = \sigma_n(\omega).$$

Let Q be the derivative matrix:

$$Q(i, j) = P'(0, i, j).$$

Let

$$q(i) = -Q(i, i).$$

Let

$$\Gamma(i, j) = Q(i, j)/q(i) \quad \text{for } i \neq j$$
$$= 0 \quad \text{for } i = j.$$

I say

(161) $$1/q(1) = m\{d(1)\}[d(1) - d(2)].$$

And for $n \geqq 2$,

(162) $$1/q(n) = m\{d(n)\}\frac{[d(n-1) - d(n)][d(n) - d(n+1)]}{d(n-1) - d(n+1)}.$$

Indeed, (149) and (MC, 7.4) identify $1/q(n)$ as the $B_{d(n)}$-mean holding time of Y in $d(n)$. Suppose $n \geqq 2$. Then

$$1/q(n) = E_{d(n)}\{M(\sigma_n)\} \quad \text{by (160)}$$
$$= m\{d(n)\}\frac{[d(n-1) - d(n)][d(n) - d(n+1)]}{d(n-1) - d(n+1)} \quad \text{by (97).}$$

The argument for $n = 1$ is similar: approximate σ_1 by $\tau_{[d(2),N]}$ for large N.

I say:

(163) $$\Gamma(1, 2) = 1 \quad \text{and} \quad \Gamma(1, j) = 0 \quad \text{for } j > 2;$$

(164) $$\Gamma(i, j) = 0 \quad \text{for } i \geqq 2 \quad \text{and} \quad |i - j| \neq 1;$$

for $i \geqq 2$,

(165) $$\Gamma(i, i-1) = \frac{d(i) - d(i+1)}{d(i-1) - d(i+1)} \text{ and } \Gamma(i, i+1) = \frac{d(i-1) - d(i)}{d(i-1) - d(i+1)}.$$

I will only compute $\Gamma(i, i-1)$ for $i \geqq 2$. Theorems (MC, 7.21) and (149) identify $\Gamma(i, i-1)$ as the $B_{d(i)}$-probability that Y jumps to $d(i-1)$ on first leaving $d(i)$. By (160),

$$\Gamma(i, i-1) = B_{d(i)}\{X(\sigma_i) = d(i-1)\} = \frac{d(i) - d(i+1)}{d(i-1) - d(i+1)}:$$

because $\{B_x\}$ is in natural scale.

I claim:

(166) There is exactly one function h which is 1 at 1 and harmonic for Γ, namely, $h \equiv 1$.

I remind you that h is harmonic for Γ means

$$\Sigma_j \Gamma(i, j)h(j) = h(i) \quad \text{for all } i.$$

Indeed, put $i = 1$ and use (163) to get $h(2) = 1$. If $i \geqq 2$, use (164) to get

$$\Gamma(i, i-1)h(i-1) + \Gamma(i, i+1)h(i+1) = h(i).$$

Use (165) to check $0 < \Gamma(i, i-1) < 1$ and $\Gamma(i, i+1) = 1 - \Gamma(i, i-1)$. Then $h(i+1) = 1$ by induction on i.

I claim:

(167) There is exactly one measure μ which is 1 at 1 and which is invariant for Γ, namely,

$$\mu\{n\} = \frac{d(1) - d(2)}{d(n-1) - d(n)} \frac{d(n-1) - d(n+1)}{d(n) - d(n+1)} \quad \text{for } n \geqq 2.$$

I remind you that μ is invariant for Γ means

$$\Sigma_i\, \mu(i)\Gamma(i, j) = \mu(j) \quad \text{for all } j.$$

Indeed, (163–164) show that μ can be computed recursively. Using (165), check that my proposed μ is invariant.

The first atrocity is the complexity (156) of U_0, the set of times at which Y is in its adjoined state. This is particularly outrageous, because Y can reach any state from any other state in a finite number of jumps by (165); and all states are recurrent. Moreover, the "entrance and exit boundaries" of the jump process, are completely trivial by (166–167), and cannot be used to explain the complexity of $U_0(\omega)$. This possibility was first brought out by Chung (1963, 1966) and Reuter (1959). In Chung's terminology, Y has a sticky atom. The exit boundary was discussed in Chapter 4 of *MC*. The entrance boundary is the exit boundary of the reversed chain, and is related to the invariant measures like the exit boundary is to the harmonic functions.

As a preliminary to the second atrocity story, remember (*MC*, Sec. 1.3) that the coordinate process $\{\xi_n\}$ is Markov with stationary transitions Γ and starting state d, relative to the probability Γ_d. Let $k > j$, so $d(k) < d(j)$. I claim

(168) $$\Gamma_{d(k)}\{\xi_n = d(j) \text{ for some } n\} = d(k)/d(j).$$

Indeed, (149) and (*MC*, 7.33) identify the left side of (168) as the $B_{d(k)}$-probability that Y visits $d(j)$ before 0. By (152), this coincides with the $B_{d(k)}$-probability that X visits $d(j)$ before 0. Because $\{B_x\}$ is in natural scale, this visiting probability is $d(k)/d(j)$.

Let $(\Omega, \mathscr{F}, \mathscr{P})$ be a probability space, and let

$$\{W(n) : n = \ldots, -1, 0, 1, \ldots\}$$

be a D-valued stochastic process on $(\Omega, \mathscr{F})$. Let $D_j = \{d(1), \ldots, d(j)\}$. Let H_j be the set of $\omega \in \Omega$ for which $W(n, \omega) \in D_j$ for some n. Suppose there is a least such n; call it $\tau_j(\omega)$. Let G be a stochastic matrix on D. Then W is a *strongly approximate G-chain* iff for all N and $i_0, \ldots, i_N \in D$,

$$\mathscr{P}\{H_j \text{ and } W(\tau_j + n) = i_n \text{ for } 0 \leqq n \leqq N\}$$
$$= \mathscr{P}\{H_j \text{ and } W(\tau_j) = i_0\} \prod_{n=0}^{N-1} G(i_n, i_{n+1}).$$

This is a little more restrictive than the definition of (Hunt, 1960), since $\mathscr{P}$ is a probability.

I will argue

(169) There is no strongly approximate Γ-chain.

For suppose W were such an object. To begin with,

(170) Given H_j and $W(\tau_j) = d(k)$, the conditional $\mathscr{P}$-distribution of $W(\tau_j + \cdot)$ is $\Gamma_{d(k)}$.

I claim

(171) $$\mathscr{P}\{H_j \text{ and } W(\tau_j) \neq d(j)\} = 0.$$

Indeed, fix $\varepsilon > 0$. Choose i so large that

$$\mathscr{P}\{H_i \text{ and } W(\tau_i) < d(j)\} \geqq 1 - \varepsilon.$$

Given H_i and $W(\tau_i) = d(k) < d(j)$, the conditional $\mathscr{P}$-probability that $W(\tau_i + \cdot)$ first hits D_j at a place different from $d(j)$ is 0, by (170) and (1.16). On $\{H_i \text{ and } W(\tau_i) = d(k) < d(j)\}$, the place where $W(\tau_i + \cdot)$ first hits D_j coincides with the place where W first hits D_j. So

$$\mathscr{P}\{H_j \text{ and } W(\tau_j) \neq d(j)\} \leqq \varepsilon.$$

This proves (171). Let $k > j$. Then up to a null set,

$$H_j = H_k \cap \{W(\tau_k + \cdot) \text{ visits } d(j)\}.$$

Use (170–171) and (168):

$$\begin{aligned}\mathscr{P}\{H_j\} &= \mathscr{P}\{H_k\} \cdot \Gamma_{d(k)}\{\xi_n = d(j) \text{ for some } n\}\\ &= \mathscr{P}\{H_k\} \cdot d(k)/d(j)\\ &\leqq d(k)/d(j).\end{aligned}$$

Let $k \to \infty$ to see $\mathscr{P}\{H_j\} = 0$. But H_j increases to Ω as j increases to ∞, a contradiction. This proves (169).

The second atrocity is (169). To make this more vivid, let τ be the first time Y visits $d(1)$ after visiting 0. Let $W_0, W_1, \ldots$ be the succession of states visited by Y after τ, before the subsequent 0. Let $\ldots, W_{-2}, W_{-1}$ be the succession of states visited by Y before τ, but after the antecedent 0. Then

$$(\ldots, W_{-1}, W_0, W_1, \ldots) \text{ is not a strongly approximate chain.}$$

For if it were, strong Markov would make it a strongly approximate Γ-chain, contradicting (169). You should compare this with Sections 6.3 and 8.5–8.6 of *MC*.

I think you can argue that the transitions P are uniquely determined by the generator Q and condition (157) on a P-chain Y with regular sample functions. Condition (157) is equivalent to the requirement that P satisfies the forward equation (*MC*, 7.64): use (*MC*, 7.58 and 7.69). If so, how can you construct Y without reference to Brownian motion? Here is my semiformal program. Consider the maximal intervals J complementary to

$$U_0 = \{t : Y(t) = 0\},$$

and to the right of the first 0. Call J an *n-interval* iff the maximum of Y on J is $d(n)$. Then the fragments: Y to the first 0; and Y on the mth n-interval are all mutually independent. The holding times in Y can be filled in independently from (161–162), after the order of states is decided. The order of states to the first 0 is a Γ-chain. The distribution of the order of states on the mth n-interval does not depend on m; it can be generated as follows. Begin the construction by building a Γ-chain γ_0 starting from $d(n)$, conditioned on not reaching $d(n - 1)$. So γ_0 is an infinite sequence of elements of D, which starts at $d(n)$ and is always $d(n)$ or less. For $m = 1, 2, \ldots$ build a Γ-chain γ_m which starts at $d(n + m)$, and is conditioned on reaching $d(n + m - 1)$; stop γ_m just before it first hits $d(n + m - 1)$. So γ_m is a finite sequence of elements of D, which starts and ends at $d(n + m)$, and is always $d(n + m)$ or less. It may be that γ_m consists of the single entry $d(n + m)$. Make $\gamma_0, \gamma_1, \ldots$ independent. The desired order of states on an n-interval is the doubly infinite state-sequence

$$\cdots \gamma_2 \gamma_1 \gamma_0 .$$

The last problem is to describe the order in which the n-intervals occur. Remember that Z is geometric with parameter p iff

$$\text{Prob}\{Z = n\} = (1 - p)p^n \quad \text{for } n = 0, 1, \ldots .$$

Let Z_0 be the number of $(N + 1)$-intervals before the first $(1, \ldots, N)$-interval; let Z_n be the number of $(N + 1)$-intervals between the nth and the $n + 1$st $(1, \ldots, N)$-interval. I will argue

(172) $Z_0, Z_1, \ldots$ are independent and geometric, with common parameter

$$1 - \frac{d(N + 1)}{d(N)}, \qquad \text{relative to } B_0 .$$

Remember that B_x makes X Brownian motion starting from x. Let $\sigma_0 = 0$. Let τ_n be the least $t > \sigma_n$ with $X(t) = d(N + 1)$. Let σ_{n+1} be the least $t > \tau_n$

with $X(t) = 0$. By strong Markov, the fragments

$$\{X(t + \sigma_{n-1}) : 0 \leqq t \leqq \sigma_n - \sigma_{n-1}\}$$

are B_0-independent and identically distributed for $n = 1, 2, \ldots$. Let

$$M_n = \max \{X(t) : \sigma_{n-1} \leqq t \leqq \sigma_n\},$$

and let

$$L_n \text{ be the smallest } m \text{ with } d(m) \leqq M_n.$$

Then

(173a) $L_1, L_2, \ldots$ are B_0-independent and identically distributed, taking the values $m = 1, \ldots, N + 1$.

I say:

(173b) $$B_0\{L_1 = 1\} = \frac{d(N+1)}{d(1)};$$

and for $m = 2, \ldots, N + 1$,

(173c) $$B_0\{L_1 = m\} = \frac{d(N+1)}{d(m)}\left[1 - \frac{d(m)}{d(m-1)}\right].$$

I will argue (173c). The event $\{X(0) = 0 \text{ and } L_1 = m\}$ coincides with the event that X reaches $d(N + 1)$; and then visits $d(m)$ before visiting 0; and then visits 0 before visiting $d(m - 1)$. Use two doses of strong Markov and natural scale to get (173c). Return to the Y constructed from Brownian motion. Then $L_1, L_2, \ldots$ gives the indices of the n-intervals of Y in order of occurrence, for $n = 1, \ldots, N + 1$. So (173) gives (172): the stretches of $(L_1, L_2, \ldots)$ up to but not including the first visit to $(1, \ldots, N)$, and between successive visits to $(1, \ldots, N)$ are independent and geometric, with the right parameter, by strong Markov on L.

Does a similar construction work when the jump chain has only one invariant measure, only one harmonic function, and infinite capacity?

14. A GENERALIZATION

Let K be a nonempty closed subset of $(-\infty, \infty)$. Let $a = \inf K$ and $b = \sup K$, so $-\infty \leqq a \leqq b \leqq \infty$. Let K_0 be K, or $K \setminus \{a\}$, or $K \setminus \{b\}$, or $K \setminus \{a, b\}$. Give K_0 the relative topology. There is a natural definition of diffusion on K_0, as will now be sketched. Let Ω be the set of *regular* functions f from $[0, \infty)$ to K_0. Namely,

(174a) As a function from $[0, \infty)$ to the real line, f is continuous from the right with a finite limit from the left existing everywhere.

(174b) If f is discontinuous at t, then $f(t) = f(t+)$ is one endpoint of a complementary interval of K, and $f(t-)$ is the other endpoint of that interval.

By the usual bisection argument,

(174c) f has the intermediate value property: if $0 \leqq s < t$, and $y \in K_0$ is between $f(s)$ and $f(t)$ there is a u between s and t with $f(u) = y$; unless y is the endpoint of a complementary interval, in which case $y = f(u)$ or $f(u-)$.

Let X be the coordinate process on Ω. Define the σ-fields $\mathscr{B}(t)$ and $\mathscr{B}$ as in Section 1. Define Markov times τ and diffusions $\{P_x : x \in K_0\}$ as in Section 1. Assume $\{P_x\}$ regular in the analogous sense. The theory of Sections 2–10 seems to go over to the general case, with the obvious changes. In Section 2, the classification of points must be completely redone. Lemma (30) can be modified to say that for $f \in BC(K_0)$, and $x_n \to x$, and $t_n \downarrow t$,

$$E_{x_n}\{f[X(t_n)]\} \to E_x\{f[X(t)]\}.$$

This probably remains true for $t_n \uparrow t$, but may not be worth proving. In Section 3, the scale is defined, continuous, and nondecreasing on K_0, and has support K. Transforming to natural scale preserves the structure. In Section 4, check that T_t is a mapping of $BC(K_0)$ into itself; for convenience, extend $f \in BC(K_0)$ to $f \in BC(a, b)$ by linear interpolation across complementary intervals. Section 5 covers the general case. For speed measure in Section 6, extend e_J from $J \cap K_0$ to all of J by linear interpolation across complementary intervals. The extended function is convex. Speed measure is no longer strictly positive on (a, b); instead, it has support K. Proceed similarly with the rest of 6 and 7. Section 8 covers the general case. Section 9 is easily adapted.

This characterizes all processes obtained from Brownian motion by the recipe

$$F[X \circ T],$$

where: F is continuous and strictly increasing; and T is the right continuous inverse of M; while

$$M(t) = \frac{1}{2}\int_{-\infty}^{\infty} \lambda(y, t)m(dy);$$

and m is a measure locally finite on the open interval from the left endpoint to the right endpoint of its support. The continuity properties (174a, b) of the sample functions come from (119) and the normalization of T as right continuous.

Interesting K to think about are: $[0, 1] \cup [2, 3]$, the Cantor set, the integers, and the nonnegative integers. I think the last two cases give all birth and death processes which satisfy the forward equation.

Suppose K has a countable, dense, relatively open subset S: and m resides on S. Then Y is a Markov chain on S. The jump process consists of several birth and death processes patched together. And Y has fictitious states $K \setminus S$.

15. WEAK MARKOV PROCESSES

My object in this section is to present three examples and a guess to clarify the strong Markov property. You ought to review (*MC*, Sec. 6.6) before reading this section. At some point, you should look at Ray (1956). Let $I = (-\infty, \infty)$, and let Ω be the set of all functions from $[0, \infty)$ to I. Let

$$X(t, \omega) = \omega(t) \quad \text{for } 0 \leqq t < \infty \text{ and } \omega \in \Omega.$$

Let $\mathscr{B}$ be the smallest σ-field in Ω which makes all $X(t)$ measurable. Let $\mathscr{B}(t)$ be the smallest σ-field in Ω which makes all $X(s)$ with $s \leqq t$ measurable. Let S_t map Ω into itself:

$$(S_t\omega)(s) = \omega(t + s).$$

Check that S_t is $\mathscr{B}$-measurable.

As in (*MC*, 6.146a), call σ a *strict Markov time* iff $0 \leqq \sigma \leqq \infty$ and

$$\{\sigma \leqq t\} \in \mathscr{B}(t) \quad \text{for all } t.$$

Let $\mathscr{B}(\sigma)$ be the σ-field of all $A \in \mathscr{B}$ such that

$$A \cap \{\sigma \leqq t\} \in \mathscr{B}(t) \quad \text{for all } t.$$

Let S_σ map Ω into itself:

$$(S_\sigma\omega)(s) = \omega[\sigma(\omega) + s].$$

Check that S_σ is $\mathscr{B}$-measurable.

For each $x \in (-\infty, \infty)$, let P_x be a probability on $(\Omega, \mathscr{B})$. Review definition (1) of diffusion.

(175) Definition. $\{P_x\}$ is weak Markov *iff*:

$P_x\{X(0) = x\} = 1$ *for all* x, *and*

$P_{X(t)}$ *is a regular conditional* P_x*-distribution for* S_t *given* $\mathscr{B}(t)$, *for all* t *and* x.

$\{P_x\}$ *is* intermediate Markov *iff*:

$P_x\{X(0) = x\} = 1$ *for all* x, *and*

$P_{X(\sigma)}$ is a regular conditional P_x-distribution for S_σ given $\mathscr{B}(\sigma)$ on $\{\sigma < \infty\}$, for all x and strict Markov σ.

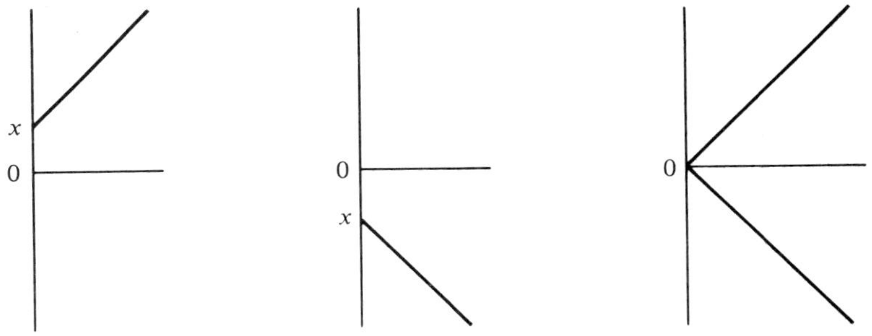

Figure 22.

(176) Example. Look at Figure 22. For $x > 0$, let P_x concentrate on the single function $t \to x + t$. For $x < 0$, let P_x concentrate on the single function $t \to x - t$. Let P_0 assign probability $\frac{1}{2}$ each to the two functions

$$t \to t \text{ and } t \to -t.$$

You check:

> $\{P_x\}$ is intermediate Markov but not a diffusion; for $\mathscr{B}(0+)$ is P_0-nontrivial, by contrast with (4).

(177) Example. Let $(\mathscr{X}, \mathscr{F})$ be the cartesian product of Borel $(0, \infty)$ and $\{-1, 1\}$. For $t \geqq 0$ and $(u, v) \in \mathscr{X}$, let

$$\begin{aligned} X(t, (u, v)) &= 0 && \text{when } 0 \leqq t \leqq u \\ &= v(t - u) && \text{when } t \geqq u. \end{aligned}$$

Then X is a process with continuous, real-valued sample functions. Let

$$\tau(u, v) = u \quad \text{and} \quad \theta(u, v) = v.$$

Let $\mathscr{P}$ be the probability on $(\mathscr{X}, \mathscr{F})$ for which:

τ and θ are independent;

τ is exponential with parameter 1;

θ is ± 1 with probability $\frac{1}{2}$ each.

For $x \geqq 0$, let P_x concentrate on the single function $t \to x + t$. For $x \leqq 0$, let P_x concentrate on the single function $t \to x - t$. Let P_0 be the $\mathscr{P}$-distribution of X. See Figure 23. I claim:

> $\{P_x\}$ is intermediate Markov but not a diffusion: because (6) fails. If σ is strict Markov, then each $\mathscr{B}(\sigma+)$-set differs from a $\mathscr{B}(\sigma)$-set only by a P_x-null set. But
>
> $$\tau = \inf\{t : X(t) \neq 0\}$$
>
> is a Markov time which is really different from any strict Markov time.

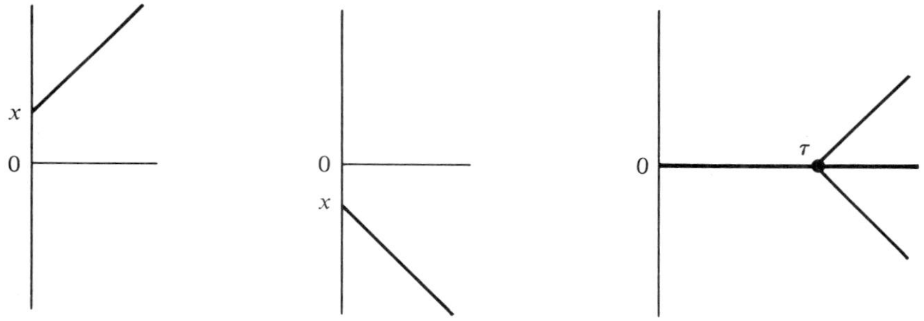

Figure 23.

PROOF. Here is the main step: the rest of the argument is up to you.

(178) Let σ be a strict Markov time on $\mathscr{X}$. Then for some real number $s = s(\sigma)$,

$$\{\sigma \leqq \tau\} \subset \{\sigma = s\}.$$

To argue (178), let $\mathscr{F}(t)$ be the σ-field in $\mathscr{X}$ generated by $X(u)$ with $u \leqq t$. Check that

$$\{\tau \geqq s\} \cap \mathscr{F}(s)$$

consists of the empty set $\emptyset$ and $\{\tau \geqq s\}$. On the other hand,

$$\{\tau < s\} \cap \mathscr{F}(s) = \{\tau < s\} \cap \mathscr{F}.$$

If $F \in \mathscr{F}(s)$, then

$$F = A \cup B,$$

where A is an $\mathscr{F}$-measurable subset of $\{\tau < s\}$, and

$$B = \emptyset \quad \text{or} \quad \{\tau \geqq s\}.$$

Fix $x \in \{\sigma \leqq \tau\}$. I only have to check $\sigma(y) \leqq \sigma(x)$, for all $y \in \{\sigma \leqq \tau\}$: because you can interchange x and y. Suppose $\sigma(x) < \infty$, and abbreviate $s = \sigma(x)$. Clearly, $\{\sigma = s\} \in \mathscr{F}(s)$. So

$$\{\sigma = s\} = A \cup B,$$

where A is an $\mathscr{F}$-measurable subset of $\{\tau < s\}$, and

$$B = \varnothing \quad \text{or} \quad \{\tau \geqq s\}.$$

But $\tau(x) \geqq \sigma(x) = s$, so $x \in B$, forcing $B = \{\tau \geqq s\}$. That is,

$$\{\sigma = s\} \supset \{\tau \geqq s\}.$$

If $y \in \{\sigma \leqq \tau\}$ and $\sigma(y) \geqq s$, then $\tau(y) \geqq \sigma(y) \geqq s$, forcing $\sigma(y) = s$. ★

For (179), let $\sigma(y, t)$ be the least $s \geqq t$ if any with $X(s) = y$, and $\sigma(y, t) = \infty$ if none. So $\sigma(y, t)$ is a strict Markov time on Ω.

(179) Example. There is a countable set D and a weak Markov system of probabilities $\{P_x\}$ such that:

(a) $P_x\{X(t) \in D\} = 1$ for all x and t;

(b) $P_x\{\sigma(y, t) < \infty\} = 1$ for all x, y, and t;

(c) P_y is not a regular conditional P_x-distribution for $S_{\sigma(y,t)}$ given $\mathscr{B}(\sigma(y, t))$, for any x, y, or positive t;

in particular, $\{P_x\}$ is not intermediate Markov;

(d) each $\mathscr{B}(\sigma+)$-set differs from a $\mathscr{B}(\sigma)$-set only by a P_x-null set, for any x and strict Markov σ.

NOTE. This should make you appreciate strong Markov (*MC*, 9.41). Here, D has an absurd topology inherited from the line. The fictitious states of X_i^n coincide for $i = 1, \ldots, 4$, because D_i is dense. In *MC*, the topology on D is discrete, and there is only the one fictitious state φ. If you pass to the star version of (*MC*, 9.17), for almost all sample functions the pieces of path taken from X_i^n will have values in $D_i \cup \{\varphi\}$ only.

CONSTRUCTION. Let D_i be countable, dense, disjoint subsets of $(-\infty, \infty)$, for $i = 1, 2, 3, 4$. Let m_i be a probability on D_i which assigns positive mass to each point of D_i. Let X_i^n be independent diffusions for $i = 1, 2, 3, 4$ and $n = 1, 2, \ldots$. Suppose X_i^n is in natural scale on $(-\infty, \infty)$, and has speed measure m_i. Let $u < v$ be points outside $D_1 \cup D_2 \cup D_3 \cup D_4 = D$. Let

Y_i^n be X_i^n up to first hitting u, for $i = 1$ or 3

Y_i^n be X_i^n up to first hitting v, for $i = 2$ or 4.

For $n \geqq 2$, make

X_1^n and X_3^n start from v

X_2^n and X_4^n start from u.

Let $x \notin D_2 \cup D_3 \cup D_4$. To construct P_x, make

$$X_1^1 \text{ start from } x$$

$$X_2^1 \text{ and } X_4^1 \text{ start from } u$$

$$X_3^1 \text{ start from } v.$$

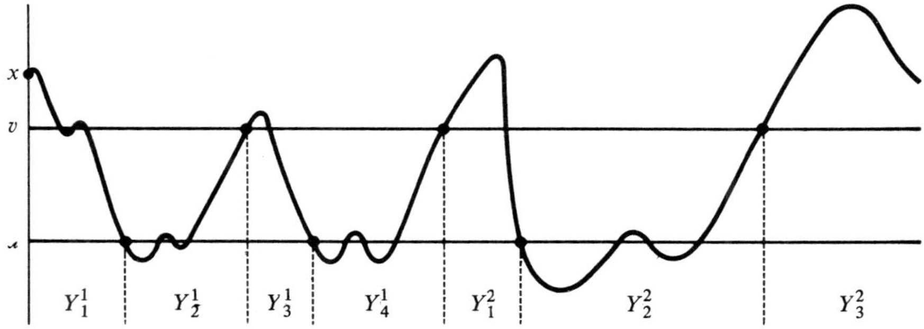

Figure 24.

Then P_x is the distribution of the process obtained as in Figure 24, by laying together the fragments

$$Y_1^1, Y_2^1, Y_3^1, Y_4^1, Y_1^2, Y_2^2, Y_3^2, Y_4^2, Y_1^3, \ldots$$

in order. Let $x \in D_2$. To construct P_x, make

$$X_2^1 \text{ start from } x$$

$$X_3^1 \text{ start from } v$$

$$X_4^1 \text{ start from } u.$$

Then P_x is the distribution of the process obtained by laying together the fragments

$$Y_2^1, Y_3^1, Y_4^1, Y_1^2, Y_2^2, Y_3^2, Y_4^2, Y_1^3, \ldots$$

in order. Let $x \in D_3$. To construct P_x, make

$$X_3^1 \text{ start from } x$$

$$X_4^1 \text{ start from } u.$$

Then P_x is the distribution of the process obtained by laying together the fragments

$$Y_3^1, Y_4^1, Y_1^2, Y_2^2, Y_3^2, Y_4^2, Y_1^3, \ldots$$

in order. Let $x \in D_4$. To construct P_x, make X_4^1 start from x. Then P_x is the distribution of the process obtained by laying together the fragments

$$Y_4^1, Y_1^2, Y_2^2, Y_3^2, Y_4^2, Y_1^3, \ldots$$

in order.

CLUE FOR WEAK MARKOV. $P_x\{X_i^n(t) \in D_i\} = 1$ by (144).

CLUE FOR PROPERTY (c). Suppose for instance that $x \notin D_2 \cup D_3 \cup D_4$. Let θ_1 be the least t with $X(t) = u$. Let θ_{n+1} be the least $t > \theta_n$ with $X(t) = u$ or v, but $X(t) \neq X(\theta_n)$. If I tell you the sample function up to $\sigma(y, t)$, you can compute the largest n with $\theta_n \leqq \sigma(y, t)$. The remainder when this n is divided by 4 really can affect your conditional opinion about the next piece of path, from $\sigma(y, t)$ until the subsequent visit to $\{u, v\}$.

I used four processes so there could be two different kinds of path before $\sigma(y, t)$, when $y \notin [u, v]$.

CLUE FOR PROPERTY (d). Use (180) below on X_i^n. ★

The next lemma improves (4).

(180) Lemma. *Suppose $\{P_x\}$ is a diffusion, and σ is strict Markov. Then each $\mathscr{B}(\sigma+)$-set differs from a $\mathscr{B}(\sigma)$-set only by a P_x-null set.*

PROOF. Let $0 < t_1 < \cdots < t_n < \infty$ and let $L_1, \ldots, L_n$ be linear Borel sets. Put $t_0 = 0$ and $t_{n+1} = \infty$. Let $A_0 = \Omega$, and for $m \geqq 1$,

$$A_m = \{X(t_v) \in L_v \text{ for } v = 1, \ldots, m\}.$$

For $m < n$ and $s < t_{m+1}$, let

$$A_m^*(s) = \{X(t_{m+v} - s) \in L_{m+v} \text{ for } v = 1, \ldots, n - m\};$$

otherwise, $A_m^*(s) = \Omega$. By strong Markov and (3.44),

$$P_x\{A_n | \mathscr{B}(\sigma+)\} = 1_{A_m} \cdot P_{X(\sigma)}\{A_m^*(\sigma)\} \quad \text{on} \quad \{t_m \leqq \sigma < t_{m+1}\},$$

at least with P_x-probability 1. So $P_x\{A_n | \mathscr{B}(\sigma+)\}$ is almost $\mathscr{B}(\sigma)$-measurable. You can extend this to get $P_x\{A | \mathscr{B}(\sigma+)\}$ almost $\mathscr{B}(\sigma)$-measurable for all $A \in \mathscr{B}$, especially $A \in \mathscr{B}(\sigma+)$. ★

Despite all this, there seems to be no real difference between Markov times and strict Markov times, for diffusions. As follows from results of P. A. Meyer,

(181) Let $\{P_x\}$ be a regular diffusion in natural scale. Let τ be a Markov time. Then there is a strict Markov time σ such that:

$$\sigma \leqq \tau \quad \text{everywhere, and}$$
$$P_x\{\sigma = \tau\} = 1 \quad \text{for all } x.$$

3

APPENDIX

1. NOTATION

★ is the end of a proof, or a discussion.

iff is if and only if.

$A \setminus B$ is the set of points in A but not in B.

$A \bigtriangleup B = (A \setminus B) \cup (B \setminus A)$.

$\emptyset$ is the empty set.

$\circ$ is composition.

If $\mathscr{F}$ is a σ-field and $A \in \mathscr{F}$, then $A\mathscr{F}$ is the σ-field of subsets of A of the form $A \cap F$, with $F \in \mathscr{F}$.

X is measurable on Y means that X is measurable relative to any σ-field over which Y is measurable. Usually, this means you can compute X in a measurable way from Y.

If π is a statement about points x, then

$$\{\pi\} = \{x : \pi(x)\} = [\pi] = [x : \pi(x)]$$

is the set of x for which $\pi(x)$ is true. And

$$\Sigma_x \{a_x : \pi(x)\} = \Sigma \{a_x : \pi(x)\}$$

is the sum of a_x over all x for which $\pi(x)$ is true.

An empty sum is 0; an empty product is 1.

$a_n = 0(b_n)$ means: there are finite K and N such that

$$|a_n| \leqq K|b_n| \quad \text{for all } n \geqq N.$$

$a_n = o(b_n)$ means: for any positive ε, there is a finite N_ε such that

$$|a_n| \leqq \varepsilon|b_n| \quad \text{for all } n \geqq N_\varepsilon.$$

I want to thank Allan Izenman for checking the final draft of this chapter.

$a_n \sim b_n$ means: there are finite, positive K and N such that

$$\frac{1}{K} b_n \leqq a_n \leqq K b_n \quad \text{for all } n \geqq N.$$

$a_n \approx b_n$ means: for any ε with $0 < \varepsilon < 1$, there is a finite N_ε such that

$$(1 - \varepsilon) b_n \leqq a_n \leqq (1 + \varepsilon) b_n \quad \text{for all } n \geqq N_\varepsilon.$$

$[0, 1) = \{x : 0 \leqq x < 1\}$.

$\langle x \rangle$ means the greatest integer $n \leqq x$.

x is *positive* means $x > 0$, while x is *nonnegative* means $x \geqq 0$. When it seems desirable, the redundancy "x is *strictly positive*" is employed. Similarly for *increasing* and *nondecreasing*.

Real-valued means in $(-\infty, \infty)$, while *extended real-valued* means in $[-\infty, \infty]$. Random variables are allowed to take infinite values without explicit warning.

Clearly usually means that the assertion which follows is clear. Sometimes, by force of habit, it means that I didn't feel like writing out the argument.

Let f be a real-valued function on $S \times T$. Then $f(s) = f(s, \cdot)$ is the real-valued function $t \to f(s, t)$ on T, while $f(t) = f(\cdot, t)$ is the real-valued function $s \to f(s, t)$ on S. Furthermore, f is used indifferently for the real-valued function $(s, t) \to f(s, t)$ on $S \times T$, the function-valued mapping $s \to f(s)$ on S, and the function-valued mapping $f \to f(t)$ on T. Whenever this threatens to get out of hand, some explanation is provided.

2. NUMBERING

In each chapter, all important formulas, definitions, theorems and so on are treated as displays and numbered consecutively from 1 on. Inside chapter a: display (b) means the display numbered b in chapter a; for $a' \neq a$, display $(a' \cdot b)$ means the display number b in chapter a′. Section $a \,.\, b$ is section b of chapter a. And $(MC, a \,.\, b)$ is the display numbered b in chapter a of MC; this kind of reference is used in *B & D* and *ACM*.

3. BIBLIOGRAPHY

(Blackwell, 1958) and Blackwell (1958) refer to the work of Blackwell listed in the bibliography with year of publication 1958. The obvious problem is settled by this device: (Lévy, 1954a). Each entry in the bibliography gives the edition I used when writing the book. In certain cases, notably (Chung, 1960), a more recent edition is now available. When this is known to me, the newer edition is referred to in parentheses following the main entry. Journals are abbreviated following *Math. Rev.* practice. I do not give references to my own articles.

This book is part of a trilogy, published by Holden-Day at San Francisco in 1970. The titles, and their abbreviations, are:

Markov Chains (*MC*)
Brownian Motion and Diffusion (*B & D*)
Approximating Countable Markov Chains (*ACM*)

4. THE ABSTRACT LEBESGUE INTEGRAL†

As usual, Ω is a set, and $\mathscr{F}$ is a σ-field of subsets of Ω; that is, $\Omega \in \mathscr{F}$ and $\mathscr{F}$ is closed under complementation and the formation of countable unions. A *probability* $\mathscr{P}$ on $\mathscr{F}$ is a countably additive, nonnegative function, with $\mathscr{P}(\Omega) = 1$. Then $(\Omega, \mathscr{F}, \mathscr{P})$ is called a *probability triple*. If $\mathscr{P}(\Omega) \leqq 1$, then $\mathscr{P}$ is a *subprobability*. On occasion, reference will be made to the *inner measure* $\mathscr{P}_*(A)$ of $A \subset \Omega$, which is $\sup\{\mathscr{P}(F): F \in \mathscr{F} \text{ and } F \subset A\}$. Similarly, the *outer measure* $\mathscr{P}^*(A) = \inf\{\mathscr{P}(F): F \in \mathscr{F} \text{ and } A \subset F\}$.

The *indicator function* 1_A is 1 on A and 0 on $\Omega \setminus A$, the set of points in Ω but not in A. A *random variable* X is an $\mathscr{F}$*-measurable* function from Ω to the real line: that is,

$$[X \leqq x] = \{X \leqq x\} = \{\omega : \omega \in \Omega \text{ and } X(\omega) \leqq x\} \in \mathscr{F}$$

for all x. If X takes infinite values, it will sometimes, but not always, be described as *extended real-valued*. A *partially defined* random variable is a random variable on $(A, A\mathscr{F})$, where $A \in \mathscr{F}$ and $A\mathscr{F} = \{A \cap B : B \in \mathscr{F}\}$. For $X \geqq 0$, the *expectation* (or $\mathscr{P}$-expectation, if several probabilities are in sight) of X is a nonnegative, extended real number $E(X)$:

$$E(1_A) = \mathscr{P}(A); \quad \text{and} \quad E(\alpha X + \beta Y) = \alpha E(X) + \beta E(Y);$$

and

(1) Monotone convergence theorem. *If* $0 \leqq X_n \uparrow X$, *then* $E(X_n) \uparrow E(X)$.

Here $\alpha \geqq 0$ and $\beta \geqq 0$; and $(\alpha X + \beta Y)(\omega) = \alpha X(\omega) + \beta Y(\omega)$; and $X_n \uparrow X$ means that $X_n(\omega)$ is nondecreasing with n and tends to $X(\omega)$ as $n \to \infty$ for each $\omega \in \Omega$. Moreover,

(2) Fatou's Lemma. *If* $X_n \geqq 0$, *then* $E(\liminf X_n) \leqq \liminf E(X_n)$.

If $\{A_n\}$ is a sequence of sets, then $\limsup A_n$ is the set of ω which are elements of infinitely many A_n; and $\liminf A_n$ is the set of ω which are

† References: (Loève, 1963, Part 1); (Neveu, 1965, Chapters 1 and 2).

elements of all but finitely many A_n. Thus,

$$1_{\liminf A_n} = \liminf 1_{A_n} \quad \text{and} \quad 1_{\limsup A_n} = \limsup 1_{A_n}.$$

Furthermore, $\Omega \setminus (\limsup A_n) = \liminf (\Omega \setminus A_n)$. By definition $A_n \to A$ iff $A = \limsup A_n = \liminf A_n$. As (2) implies,

(3) Corollary. *If $A_n \to A$, then $\mathscr{P}(A_n) \to \mathscr{P}(A)$.*

In general, $X = X^+ - X^-$, where $X^+ = X \cup 0 = \max\{X, 0\}$, and

$$E(X) = \int X\, d\mathscr{P} = \int X(\omega)\, \mathscr{P}(d\omega) = E(X^+) - E(X^-),$$

except $\infty - \infty$ is taboo. Write $E(1_A X) = \int_A X = \int_A X\, d\mathscr{P}$.

(4) Dominated convergence theorem. *If $X_n \to X$ and $|X_n| \leqq Y$ for all n and $E(Y) < \infty$, then $E(X_n) \to E(X)$.*

If $\lim_{k \to \infty} \int_{\{|X_n| \geqq k\}} |X_n|\, d\mathscr{P} = 0$ uniformly in n, the X_n are *uniformly integrable*.

(5) Criterion. *$\{X_n\}$ is uniformly integrable if either (a) or (b) holds:*

(a) *$E(|X_n|^p) \leqq K$ for all n, for some $p > 1$ and $K < \infty$.*

(b) *$\mathscr{P}\{|X_n| \geqq x\} \leqq \mathscr{P}\{Y \geqq x\}$ for all n and all $x > K$, where $0 \leqq K < \infty$ and $0 \leqq Y$ and $E(Y) < \infty$.*

(6) Theorem. *If $X_n \to X$ and the X_n are uniformly integrable, $E(X_n) \to E(X)$; in fact, $E(|X - X_n|) \to 0$.*

Easy, useful estimates:

(7) Chebychev inequality. *$\mathscr{P}\{X \geqq k\} \leqq k^{-1} E(X)$ for $X \geqq 0$ and $k > 0$.*

(8) Schwarz inequality. *$[E(XY)]^2 \leqq E(X^2) \cdot E(Y^2)$. Equality holds iff $\mathscr{P}\{Y = aX\} = 1$ for some constant a.*

(9) Jensen inequality. *Suppose $E(|X|) < \infty$ and f is convex. Then $E[f(X)] \geqq f[E(X)]$. Equality holds iff f is linear on an interval $[a, b]$ with $\mathscr{P}\{a \leqq X \leqq b\} = 1$.*

PROOF. Use (76) below. ★

Useful miscellany:

(10) Theorem. **(a)** *Let X and Y be random variables. Suppose*

$$\int_A X\, d\mathscr{P} = \int_A Y\, d\mathscr{P}$$

for all measurable A. Then $\mathscr{P}\{X = Y\} = 1$.

(b) $E(|X|) < \infty$ *implies* $\mathscr{P}\{|X| < \infty\} = 1$.
(c) $E(|X|) = 0$ *implies* $\mathscr{P}\{X = 0\} = 1$.

Absolute continuity

Suppose P and Q are two probabilities on $\mathscr{F}$. Then $P \ll Q$, or P is *absolutely continuous* with respect to Q, iff $Q(A) = 0$ implies $P(A) = 0$.

(11) Radon-Nikodym Theorem. *$P \ll Q$ iff there is a nonnegative measurable finite f with*

$$P(A) = \int_A f\,dQ.$$

This f is unique up to changes on null sets, by (10a), *and is the* Radon-Nikodym *derivative of P with respect to Q. For any nonnegative measurable g,*

$$\int g\,dP = \int gf\,dQ.$$

Say P is *orthogonal* or *singular* with respect to Q, or $P \perp Q$, iff there is a set A with $P(A) = 0$ and $Q(\Omega \setminus A) = 0$.

(12) Lebesgue decomposition. *$P = P_0 + P_1$, where $P_0 \ll Q$ and $P_1 \perp Q$ and the P_i are subprobabilities.*

Say P is equivalent to Q, or $P \equiv Q$, iff $P \ll Q$ and $Q \ll P$.

Convergence

Suppose X_n and X are finite almost surely. Say X_n *converges to* X, or $X_n \to X$, *in $\mathscr{P}$-probability* iff

$$\mathscr{P}\{|X_n - X| \geqq \varepsilon\} \to 0$$

for any $\varepsilon > 0$. Say $\{X_n\}$ is *fundamental in probability* iff

$$\lim_{n,m\to\infty} \mathscr{P}\{|X_n - X_m| \geqq \varepsilon\} = 0$$

for any $\varepsilon > 0$. For extended real-valued X_n and X, say $X_n \to X$ in $\mathscr{P}$-probability iff for any positive, finite ε and K:

$$\mathscr{P}\{|X_n - X| \geqq \varepsilon \text{ and } |X| < \infty\} \to 0$$

$$\mathscr{P}\{X_n \leqq K \text{ and } X = \infty\} \to 0$$

$$\mathscr{P}\{X_n \geqq -K \text{ and } X = -\infty\} \to 0.$$

(13) Theorem. *Suppose $\{X_n\}$ are finite almost surely and fundamental in* *probability. Then there is a random variable X, also finite almost surely, such that $X_n \to X$ in probability. Conversely, if $\{X_n\}$ and X are finite almost surely, and $X_n \to X$ in probability, then $\{X_n\}$ is fundamental in probability.*

Say X_n *converges to* X, or $X_n \to X$, *almost surely* iff

$$\mathscr{P}\{X_n \to X\} = 1.$$

(14) Theorem. *If $X_n \to X$ almost surely, then $X_n \to X$ in probability. If $X_n \to X$ in probability, there is a nonrandom subsequence n^* such that $X_{n^*} \to X$ almost surely.*

If $\pi(\omega)$ is a statement about ω, then π *almost everywhere*, or π $\mathscr{P}$-a.e. or π *almost surely* means $\mathscr{P}(\omega:\pi(\omega)$ is false$\} = 0$. Similarly, π a.e. *on A* means $\mathscr{P}\{\omega : \pi(\omega)$ is false and $\omega \in A\} = 0$. Finally, π *almost nowhere* means that $\mathscr{P}\{\omega:\pi(\omega)$ is true$\} = 0$.

The L^p-spaces

A random variable X is in L^p relative to $\mathscr{P}$ iff $\int |X|^p \, d\mathscr{P} < \infty$. The p-th root of this number is the L^p-norm of X. A sequence $X_n \to X$ in L^p if the norm of $X - X_n$ tends to 0. After identifying functions which are equal a.e., L^p is a Banach space for $p \geqq 1$. This popular fact is not used in the book.

The results of this section (except that uniform integrability gets more complicated) are usually true, and sometimes used, for measures $\mathscr{P}$ which are not probabilities; a measure on $\mathscr{F}$ is nonnegative and countably additive. In places like (11), you have to assume that $\mathscr{P}$ is σ-finite:

$$\Omega = \textstyle\bigcup_{i=1}^{\infty} \Omega_i \quad \text{with} \quad \mathscr{P}(\Omega_i) < \infty.$$

For the rest, suppose $\mathscr{P}$ is a probability; although I occasionally use converse Fubini (22) for σ-finite measures.

5. ATOMS†

If Σ is a σ-field of subsets of Ω and $\omega \in \Omega$, the Σ-*atom* containing ω is $\Sigma(\omega)$, the intersection of all the Σ-sets containing ω. Say Σ is *separable* or *countably generated* iff it is the smallest σ-field which includes some countable collection $\mathscr{C}$ of sets. In this case, let $\mathscr{A}$ be the smallest *field* containing $\mathscr{C}$. Namely, $\Omega \in \mathscr{A}$; and $\mathscr{A}$ is closed under the formation of complements and finite unions. Then $\mathscr{A}$ is countable and generates Σ. That is, Σ is the smallest σ-field which includes $\mathscr{A}$. Let $\mathscr{A}(\omega)$ be the intersection of all the $\mathscr{A}$-sets

† References: (Blackwell, 1954); (Loève, 1963, Secs. 1.6, 25.3 and 26.2).

containing ω, which by definition is the $\mathscr{A}$-atom containing ω. Then $\mathscr{A}(\omega) \in \Sigma$, and $\mathscr{A}(\omega) = \Sigma(\omega)$. Indeed, $\mathscr{A}(\omega)$ is wholly included in or wholly disjoint from each $\mathscr{A}$-set. By the *monotone class argument*, which I will make in a second, this goes for Σ as well.

Call M a *monotone class* iff:

(a) $A_n \in M$ and $A_1 \subset A_2 \subset \cdots$ imply $\bigcup_n A_n \in M$; and

(b) $A_n \in M$ and $A_1 \supset A_2 \supset \cdots$ imply $\bigcap_n A_n \in M$.

The monotone class argument. Let M be the set of $A \in \Sigma$ such that: $\mathscr{A}(\omega)$ is wholly included in or wholly disjoint from A. Then M is monotone and includes $\mathscr{A}$. Now (15) below implies $M \supset \Sigma$.

(15) Lemma. *The smallest monotone class which includes a field $\mathscr{A}$ coincides with the smallest σ-field which includes $\mathscr{A}$.*

(16) Theorem. *Let $\mathscr{C}$ be a collection of sets which is closed under intersection and generates the σ-field $\mathscr{F}$. Let P and Q be two subprobabilities on $\mathscr{F}$, which agree on $\mathscr{C}$. Suppose Ω is a countable union of pairwise disjoint elements of $\mathscr{C}$, or more generally that $P(\Omega) = Q(\Omega)$. Then $P = Q$ on $\mathscr{F}$.*

PROOF. Let $\mathscr{E}$ be the class of $A \in \mathscr{F}$ with $P(A) = Q(A)$. Clearly, $\Omega \in \mathscr{E}$ and $\mathscr{C} \subset \mathscr{E}$. If $A \in \mathscr{E}$ and $B \in \mathscr{E}$ and $A \supset B$, then $A \setminus B \in \mathscr{E}$ because $R(A \setminus B) = R(A) - R(B)$ for $R = P$ or Q. If $A \in \mathscr{E}$ and $B \in \mathscr{E}$ and $A \cap B = \emptyset$, then $A \cup B \in \mathscr{E}$, for a similar reason.

If $A_i \in \mathscr{C}$ for $i = 1, \ldots, n$, then $B_n = \bigcup_{i=1}^n A_i \in \mathscr{E}$ by induction on n. The case $n = 1$ is clear. And

$$\begin{aligned} B_{n+1} &= B_n \cup A_{n+1} \\ &= B_n \cup (A_{n+1} \setminus B_n) \\ &= B_n \cup (A_{n+1} \setminus C_n) \end{aligned}$$

where

$$C_n = A_{n+1} \cap B_n = \textstyle\bigcup_{i=1}^n (A_{n+1} \cap A_i).$$

Now $A_{n+1} \cap A_i \in \mathscr{C}$, because $\mathscr{C}$ was assumed closed under intersection. So C_n, being the union of n sets in $\mathscr{C}$, is in $\mathscr{E}$ by inductive assumption. But $C_n \subset A_{n+1}$, so $A_{n+1} \setminus C_n \in \mathscr{E}$. Finally, $A_{n+1} \setminus C_n$ is disjoint from B_n, so its union B_{n+1} with B_n is in $\mathscr{E}$. This completes the induction.

Let $A^* = A$ or $\Omega \setminus A$. If $A_i \in \mathscr{C}$ for $i = 1, \ldots, n$, I will get

$$B = \textstyle\bigcap_{i=1}^n A_i^* \in \mathscr{E}.$$

Using the assumption that $\mathscr{C}$ is closed under intersection, you can rewrite B as

$$B = [\textstyle\bigcap_{i=1}^m (\Omega \setminus C_i)] \cap D,$$

with $C_1, \ldots, C_m$ and D in $\mathscr{C}$. Let $C = \bigcup_{i=1}^m C_i$. Then

$$B = D \setminus C$$
$$= D \setminus (C \cap D).$$

Now $C \cap D \subset D$ and $C \cap D = \bigcup_{i=1}^m (C_i \cap D) \in \mathscr{E}$, because $C_i \cap D \in \mathscr{C}$. This forces $B \in \mathscr{E}$.

If $A_i \in \mathscr{C}$ for $i = 1, \ldots, n$, the field generated by $A_1, \ldots, A_n$ is included in $\mathscr{E}$; the typical atom was displayed above as $B \in \mathscr{E}$, and any nonempty set in the field is a disjoint union of some atoms. Consequently, the field $\mathscr{A}$ generated by $\mathscr{C}$ is included in $\mathscr{E}$. Of course, $\mathscr{A}$ generates $\mathscr{F}$. Now use the monotone class argument: $\mathscr{E}$ is a monotone class, and includes $\mathscr{A}$; so $\mathscr{E}$ includes the smallest monotone class which includes $\mathscr{A}$, namely $\mathscr{F}$. ★

Let Σ be a σ-field of subsets of Ω. The set of probabilities on Σ is *convex*: the convex combination of two probabilities is again a probability. A probability is *extreme* iff it cannot be represented as the convex combination of two distinct probabilities. A probability is 0–1 iff it only assumes the values 0 and 1; sometimes, Σ is called *trivial* relative to such a probability.

(17) Theorem. *Let m be a probability on Σ.*

(a) *m is extreme iff m is 0–1.*

(b) *Suppose Σ is countably generated. Then m is 0–1 iff $m(B) = 1$ for some atom B of Σ.*

PROOF. *Claim (a).* Suppose m is not 0–1. Then $0 < m(A) < 1$ for some $A \in \Sigma$. And

$$m = m(A) \cdot m(\cdot \mid A) + [1 - m(A)] \cdot m(\cdot \mid \Omega \setminus A)$$

is not extreme. Conversely, suppose m is not extreme. Then

$$m = pm' + (1 - p)m''$$

for $0 < p < 1$ and $m' \neq m''$. Find $A \in \Sigma$ with $m'(A) \neq m''(A)$. Conclude $0 < m(A) < 1$ and m is not 0-1.

Claim (b). The *if* part is easy. For *only if*, let $\mathscr{A}$ be a countable generating field for Σ. Let $\mathscr{A}_i$ be the set of $A \in \mathscr{A}$ such that $m(A) = i$, for $i = 0$ or 1. Now $\mathscr{A} = \mathscr{A}_0 \cup \mathscr{A}_1$, and $A \in \mathscr{A}_0$ iff $\Omega \setminus A \in \mathscr{A}_1$. Let B be the intersection of all $A \in \mathscr{A}_1$. Then $B \in \Sigma$ has m-probability 1, and in particular is nonempty. Fix $\omega \in B$. If $A \in \mathscr{A}_1$, then $A \supset B$. If $A \in \mathscr{A}_0$, then $\Omega \setminus A \in \mathscr{A}_1$ and $A \cap B = \emptyset$. Thus, $\omega \in A \in \mathscr{A}$ iff $A \in \mathscr{A}_1$, and B is an atom. ★

Say X is a *measurable mapping* from $(\Omega, \mathscr{F})$ to $(\Omega', \mathscr{F}')$ iff X is a function from Ω to Ω' and

$$X^{-1}\mathscr{F}' = \{X^{-1}A : A \in \mathscr{F}'\} \subset \mathscr{F}.$$

(18) Theorem. *Let X be a measurable mapping from $(\Omega, \mathscr{F})$ to $(\Omega', \mathscr{F}')$. If $\mathscr{F}'$ is countably generated, so is $X^{-1}\mathscr{F}'$. The atoms of $X^{-1}\mathscr{F}'$ are precisely the X-inverse images of the atoms of $\mathscr{F}'$.*

6. INDEPENDENCE†

Let $(\Omega, \mathscr{F}, \mathscr{P})$ be a probability triple. Sub-σ-fields $\mathscr{F}_1, \mathscr{F}_2, \ldots$ are *independent* (with respect to $\mathscr{P}$) iff $A_i \in \mathscr{F}_i$ implies

$$\mathscr{P}(A_1 \cap A_2 \cap \cdots) = \mathscr{P}(A_1) \cdot \mathscr{P}(A_2) \cdots.$$

Random variables $X_1, X_2, \ldots$ are *independent* iff the σ-fields they *span* are independent; the σ-field *spanned* or *generated* by X_i is the smallest σ-field with respect to which X_i is measurable. Sets $A, B, \ldots$ are *independent* iff $1_A, 1_B, \ldots$ are independent.

(19) Borel-Cantelli Lemma. **(a)** $\Sigma\mathscr{P}\{A_n\} < \infty$ *implies* $\mathscr{P}\{\limsup A_n\} = 0$;

(b) $\Sigma\mathscr{P}\{A_n\} = \infty$ *and* $A_1, A_2, \ldots$ *independent implies* $\mathscr{P}\{\limsup A_n\} = 1$.

Suppose X_i is a *measurable mapping* from $(\Omega, \mathscr{F})$ to $(\Omega_i, \mathscr{F}_i)$ for $i = 1, 2$. The *distribution* or $\mathscr{P}$-*distribution* $\mathscr{P}X_1^{-1}$ of X_1 is a probability on $\mathscr{F}_1$: namely,

$$(\mathscr{P}X_1^{-1})(A) = \mathscr{P}(X_1^{-1}A) \quad \text{for } A \in \mathscr{F}_1.$$

(20) Change of variables formula. *If f is a random variable on $(\Omega_1, \mathscr{F}_1)$, then*

$$E[f(X_1)] = \int_{x_1 \in \Omega_1} f(x_1)(\mathscr{P}X_1^{-1})(dx_1).$$

Let X_1 and X_2 be independent: that is, $X_1^{-1}\mathscr{F}_1$ and $X_2^{-1}\mathscr{F}_2$ are. Let $\mathscr{F}_1 \times \mathscr{F}_2$ be the smallest σ-field of $\Omega_1 \times \Omega_2$ containing all sets $A_1 \times A_2$ with $A_i \in \mathscr{F}_i$. Let f be a random variable on $(\Omega_1 \times \Omega_2, \mathscr{F}_1 \times \mathscr{F}_2)$ such that $E[f(X_1, X_2)]$ exists.

(21) Fubini's theorem. *If X_1 and X_2 are independent,*

$$\begin{aligned} E[f(X_1, X_2)] &= \int_{x_1 \in \Omega_1} E[f(x_1, X_2)]\, \mathscr{P}X_1^{-1}(dx_1) \\ &= \int_{x_2 \in \Omega_2} E[f(X_1, x_2)]\, \mathscr{P}X_2^{-1}(dx_2). \end{aligned}$$

In particular,

(21a) $\quad E(X_1 X_2) = E(X_1) \cdot E(X_2).$

† References: (Loève, 1963, Sections 8.2 and 15); (Neveu, 1965, Section IV.4).

Conversely, suppose $(\Omega_i, \mathscr{F}_i, \mathscr{P}_i)$ are probability triples for $i = 1, 2$. Let $\Omega = \Omega_1 \times \Omega_2$ and $\mathscr{F} = \mathscr{F}_1 \times \mathscr{F}_2$.

(22) Converse Fubini. *There is a unique probability* $\mathscr{P} = \mathscr{P}_1 \times \mathscr{P}_2$ *on* $\mathscr{F}_1 \times \mathscr{F}_2$ *such that*

$$\mathscr{P}(A_1 \times A_2) = \mathscr{P}_1(A_1) \cdot \mathscr{P}_2(A_2)$$

for $A_i \in \mathscr{F}_i$. *Then*

$$\int f \, d\mathscr{P} = \int \int f(x_1, x_2) \, \mathscr{P}_1(dx_1) \, \mathscr{P}_2(dx_2)$$

for nonnegative $(\mathscr{F}_1 \times \mathscr{F}_2)$*-measurable* f.

PROOF. The uniqueness comes from (16). For existence, if $A \in \mathscr{F}_1 \times \mathscr{F}_2$, let $A(\omega_1)$ be the ω_1*-section* of A, namely, the set of $\omega_2 \in \Omega_2$ with $(\omega_1, \omega_2) \in A$. Let

$$\mathscr{P}\{A\} = \int \mathscr{P}_2\{A(\omega_1)\} \, \mathscr{P}_1(d\omega_1). \qquad \bigstar$$

Let $X_1, X_2, \ldots$ be independent and identically distributed. Suppose X_1 has finite mean μ.

(23a) Weak law of large numbers. $n^{-1}(X_1 + \cdots + X_n)$ *converges to* μ *in probability*.

(23b) Strong law of large numbers. $n^{-1}(X_1 + \cdots + X_n)$ *converges to* μ *with probability* 1.

Let $\mathscr{F}_1, \mathscr{F}_2, \ldots$ be independent sub-σ-fields in $(\Omega, \mathscr{F}, \mathscr{P})$. Let $\mathscr{F}^{(n)}$ be the σ-field generated by $\mathscr{F}_n, \mathscr{F}_{n+1}, \ldots$. The *tail* σ*-field* $\mathscr{F}^{(\infty)}$ is $\bigcap_{n=1}^{\infty} \mathscr{F}^{(n)}$.

(24) Kolmogorov 0-1 Law. *Each* $\mathscr{F}^{(\infty)}$*-set has* $\mathscr{P}$*-probability* 0 *or* 1.

7. CONDITIONING†

Let $\mathscr{A}$ be a sub-σ-field of $\mathscr{F}$, and let X be a random variable with expectation. The *conditional expectation* or $\mathscr{P}$*-expectation of* X *given* $\mathscr{A}$ is $E\{X|\mathscr{A}\} = Y$, the $\mathscr{A}$-measurable random variable Y such that $\int_A Y \, d\mathscr{P} = \int_A X \, d\mathscr{P}$ for all $A \in \mathscr{A}$. For $B \in \mathscr{F}$, the *conditional probability* or $\mathscr{P}$*-probability of* B *given* $\mathscr{A}$ is $\mathscr{P}\{B|\mathscr{A}\} = E\{1_B|\mathscr{A}\}$. If Z is a measurable mapping, $E\{X|Z\} = E\{X|\mathscr{A}\}$, where $\mathscr{A}$ is the σ-field spanned by Z. According to convenience, { } changes to [ ] or ( ). Conditional expectations are unique up to changes on sets of measure 0, by (10a), and exist by Radon–Nikodym (11).

† References: (Loève, 1963, Sections 24 and 25); (Neveu, 1965, Chapter IV).

Let $\mathscr{B}$ be a sub-σ-field of $\mathscr{A}$. These facts about conditional expectations are used rather casually: equality and inequality are only a.e.

(25) Facts. **(a)** *$X \geqq 0$ implies $E\{X|\mathscr{A}\} \geqq 0$.*

(b) *$E\{X|\mathscr{A}\}$ depends linearly on X.*

(c) *$E\{X|\mathscr{A}\} = X$ if X is $\mathscr{A}$-measurable.*

(d) *$E\{XY|\mathscr{A}\} = XE\{Y|\mathscr{A}\}$ if X is $\mathscr{A}$-measurable.*

(e) *$E\{X\} = E\{E(X|\mathscr{A})\}$.*

(f) *$E\{E(X|\mathscr{A})|\mathscr{B}\} = E\{X|\mathscr{B}\}$.*

(g) *$E\{X|\mathscr{A}\} = E\{X|\mathscr{B}\}$ if $E\{X|\mathscr{A}\}$ is $\mathscr{B}$-measurable.*

(h) *If X is independent of $\mathscr{A}$, then $E\{X|\mathscr{A}\} = E\{X\}$.*

Say $\mathscr{A}$ is trivial *iff $\mathscr{P}(A) = 0$ or 1 for any $A \in \mathscr{A}$.*

(i) *If $\mathscr{A}$ is trivial, then $\mathscr{A}$ is independent of X, and $E\{X|\mathscr{A}\} = E\{X\}$.*

8. MARTINGALES†

Let T be a subset of the line. For $t \in T$, let $\mathscr{F}_t$ be a sub-σ-field of $\mathscr{F}$, and let X_t be an $\mathscr{F}_t$-measurable function on Ω, with finite expectation. Suppose that $s < t$ in T implies: $\mathscr{F}_s \subset \mathscr{F}_t$, and for $A \in \mathscr{F}_s$,

$$\int_A X_s \, d\mathscr{P} = \int_A X_t \, d\mathscr{P}.$$

Then $\{X_t, \mathscr{F}_t : t \in T\}$ is a *martingale*, or $\{X_t\}$ is a *martingale relative* to $\{\mathscr{F}_t\}$. If under similar circumstances,

$$\int_A X_s \, d\mathscr{P} \geqq \int_A X_t \, d\mathscr{P},$$

then $\{X_t\}$ is an *expectation-decreasing* martingale relative to $\{\mathscr{F}_t\}$. If $\{\mathscr{F}_t\}$ is not specified, then $\mathscr{F}_t$ is the σ-field generated by X_s for $s \in T$ with $s \leqq t$.

(26) Example. Suppose $\mathscr{F}_t$ is a sub-σ-field of $\mathscr{F}$ for each $t \in T$, and $\mathscr{F}_s \subset \mathscr{F}_t$ for $s < t$. Let X be a random variable on $(\Omega, \mathscr{F})$ with finite expectation, and $X_t = E\{X|\mathscr{F}_t\}$. Then $\{X_t, \mathscr{F}_t : t \in T\}$ is a martingale.

(27) Lemma. *Let $\{X_t\}$ be an expectation-decreasing martingale, and let f be concave and nondecreasing. Then $\{f(X_t)\}$ is an expectation-decreasing martingale.*

† References: (Doob, 1953, Chapter VII, Sections 1–4); (Loève, 1963, Section 29); (Neveu, 1965, Section IV.5).

Suppose $T = \{0, 1, \dots\}$. Let τ be a random variable on $(\Omega, \mathscr{F}, \mathscr{P})$, which is ∞ or is T. Suppose $\{\tau = t\} \in \mathscr{F}_t$ for all $t \in T$. Then τ is *admissible*, or a *stopping time*. The *pre-τ* σ-field $\mathscr{F}_\tau$ is the σ-field of $A \in \mathscr{F}$ such that $A \cap \{\tau = t\} \in \mathscr{F}_t$ for all $t \in T$.

(28) Theorem. *Suppose* $T = \{0, 1, \dots\}$. *Suppose* $\{X_t, \mathscr{F}_t\colon\ t \in T\}$ *is a martingale, and* τ *is admissible.*

(i) $\displaystyle\int_{\{A \text{ and } \tau \leqq n\}} X_\tau \, d\mathscr{P} = \int_{\{A \text{ and } \tau \leqq n\}} X_n \, d\mathscr{P}$ *for all* $A \in \mathscr{F}_\tau$.

(ii) *For each* n, *the variables* $Z_\tau = X_\tau \cdot 1_{\{\tau \leqq n\}}$ *are uniformly integrable, as* τ *varies. In fact, for any fixed* $M > 0$,

$$\int_{\{|Z_\tau| > k\}} |Z_\tau| \, d\mathscr{P} \leqq k^{-1} M E(|X_n|) + \int_{\{|X_n| > M\}} |X_n| \, d\mathscr{P}.$$

Suppose $\mathscr{P}\{\tau < \infty\} = 1$, *and*

(a) $E(|X_\tau|) < \infty$

(b) $\liminf_{n \to \infty} \displaystyle\int_{\{\tau > n\}} |X_n| \, d\mathscr{P} = 0.$

Then (X_0, X_τ) *is a martingale, so*

(iii) $E(X_0) = E(X_\tau)$.

PROOF. Let $A \in \mathscr{F}_\tau$. Let $A_m = \{A \text{ and } \tau = m\}$. Then $A_m \in \mathscr{F}_m$. Fix n. If $m \leqq n$,

$$\int_{A_m} X_\tau \, d\mathscr{P} = \int_{A_m} X_m \, d\mathscr{P} = \int_{A_m} X_n \, d\mathscr{P}.$$

Sum out $m = 0, \dots, n$ to prove (i). For (ii), put $A = \{X_\tau \geqq 0\}$ or $\{X_\tau < 0\}$ to see $E(|Z_\tau|) \leqq E(|X_n|)$, so

$$\mathscr{P}\{|Z_\tau| > k\} < k^{-1} E(|X_n|).$$

Then put $A = \{X_\tau > k\}$ or $\{X_\tau < -k\}$ to see

$$\begin{aligned}\int_{\{|Z_\tau| > k\}} |Z_\tau| \, d\mathscr{P} &\leqq \int_{\{|Z_\tau| > k\}} |X_n| \, d\mathscr{P} \\ &\leqq M\mathscr{P}\{|Z_\tau| > k\} + \int_{\{|X_n| > M\}} |X_n| \, d\mathscr{P}.\end{aligned}$$

For (iii), let $A \in \mathscr{F}_0$. Let $A_m = \{A \text{ and } \tau = m\}$. Then $A_m \in \mathscr{F}_m$. If $m \leqq n$,

$$\int_{A_m} X_\tau \, d\mathscr{P} = \int_{A_m} X_m \, d\mathscr{P} = \int_{A_m} X_n \, d\mathscr{P}.$$

Sum out $m = 0, \ldots, n$:

$$\int_{\{A \text{ and } \tau \leqq n\}} X_\tau \, d\mathscr{P} = \int_{\{A \text{ and } \tau \leqq n\}} X_n \, d\mathscr{P}.$$

Now

$$\int_A X_0 \, d\mathscr{P} = \int_A X_n \, d\mathscr{P}$$

$$= \int_{\{A \text{ and } \tau \leqq n\}} X_n \, d\mathscr{P} + \int_{\{A \text{ and } \tau > n\}} X_n \, d\mathscr{P}$$

$$= \int_{\{A \text{ and } \tau \leqq n\}} X_\tau \, d\mathscr{P} + \int_{\{A \text{ and } \tau > n\}} X_n \, d\mathscr{P}.$$

This doesn't use (a–b). By (a),

$$\lim_{n \to \infty} \int_{\{A \text{ and } \tau \leqq n\}} X_\tau \, d\mathscr{P} = \int_A X_\tau \, d\mathscr{P}.$$

By (b), if n increases through the right subsequence,

$$\int_{\{A \text{ and } \tau > n\}} X_n \, d\mathscr{P} \to 0.$$ ★

EXAMPLE. Let the Y_n be independent and identically distributed, each being 0 or 2 with probability $\frac{1}{2}$ each. Let $X_0 = 1$ and let $X_n = Y_1 \cdots Y_n$ for $n \geqq 1$. Let τ be the least n if any with $X_n = 0$, and $\tau = \infty$ if none. Then $\{X_n\}$ is a nonnegative martingale, τ is a stopping time, $E(\tau) < \infty$, and $X_\tau = 0$ almost surely. This martingale was proposed by David Gilat. ★

EXAMPLE. There is a martingale, and a stopping time with finite mean, which satisfy (28b) but not (28a).

DISCUSSION. For $n \geqq 1$, let

$$a_n = \frac{(n-1)^2}{\log(3+n)} \quad \text{and} \quad b_{n+1} = a_{n+1}^2 - a_n^2.$$

Let $b_1 = 1$. Check $b_n > 0$ for all n. Let Y_n be $N(0, b_n)$, and let $Y_1, Y_2, \ldots$ be independent. Let $X_n = Y_1 + \cdots + Y_n$, so X_n is a martingale. For $n \geqq 1$, let

$$p_n = \left(\frac{1}{n}\right)^2 - \left(\frac{1}{n+1}\right)^2,$$

so

$$\Sigma_{m=n}^{\infty} p_m = \left(\frac{1}{n}\right)^2.$$

Let f be a measurable function from $(-\infty, \infty)$ to $\{1, 2, \ldots\}$, with

$$\mathscr{P}\{f(Y_1) = n\} = p_n \quad \text{for } n = 1, 2, \ldots.$$

So $\tau = f(Y_1)$ is a stopping time, with finite mean.

Let θ be the distribution of Y_1, and let $S_n = Y_2 + \cdots + Y_n$, so $S_1 = 0$. Now Y_1 is independent of $S_1, S_2, \ldots$, and $X_n = Y_1 + S_n$, and S_n is $N(0, a_n^2)$. Compute.

$$\begin{aligned} E\{|X_\tau|\} &= \int E\{|y + S_{f(y)}|\}\, \theta(dy) \\ &\geqq \int E\{|S_{f(y)}|\}\, \theta(dy) \\ &= c \int a_{f(y)}\, \theta(dy) \qquad \text{where } c = E\{|Y_1|\} \\ &= c\, \Sigma_{n=1}^{\infty}\, p_n a_n \\ &= \infty. \end{aligned}$$

Continuing,

$$\begin{aligned} \int_{\{\tau > n\}} |X_n|\, d\mathscr{P} &= \int_{\{f > n\}} E\{|y + S_n|\}\, \theta(dy) \\ &\leqq \int_{\{f > n\}} [|y| + E\{|S_n|\}]\, \theta(dy) \\ &= \int_{\{f > n\}} |y|\, \theta(dy) + c a_n \cdot \theta\{f > n\} \\ &\to 0. \end{aligned}$$

★

Suppose $T = [0, \infty)$. Let τ be a random variable on $(\Omega, \mathscr{F}, \mathscr{P})$, which is ∞ or in T. Suppose $\{\tau < t\} \in \mathscr{F}_t$ for all $t \in T$. Then τ is *admissible*, or a *stopping time*.

(29) Theorem. *Suppose $T = [0, \infty)$. Suppose $\{X_t, \mathscr{F}_t : t \in T\}$ is a martingale, and τ is admissible. Suppose*

(a) *$t \to X_t(\omega)$ is continuous from the right, for each ω*

(b) *$E(|X_\tau|) < \infty$*

(c) *$\liminf_{t\to\infty} \int_{\{\tau > t\}} |X_t|\, d\mathscr{P} = 0$.*

Then (X_0, X_τ) is a martingale, so $E(X_0) = E(X_\tau)$.

PROOF. Fix t. Let τ_n be the least $(j/n)t \geqq \tau$. Fix $A \in \mathscr{F}_0$. As in (28),

$$\int_{\{A \text{ and } \tau \leqq t\}} X_{\tau_n}\, d\mathscr{P} = \int_{\{A \text{ and } \tau \leqq t\}} X_t\, d\mathscr{P}.$$

Let $n \to \infty$; use (a) and uniform integrability from (28):

$$\int_{\{A \text{ and } \tau \leqq t\}} X_\tau\, d\mathscr{P} = \int_{\{A \text{ and } \tau \leqq t\}} X_t\, d\mathscr{P}.$$

Now

$$\begin{aligned}\int_A X_0\, d\mathscr{P} &= \int_A X_t\, d\mathscr{P}\\ &= \int_{\{A \text{ and } \tau \leqq t\}} X_t\, d\mathscr{P} + \int_{\{A \text{ and } \tau > t\}} X_t\, d\mathscr{P}\\ &= \int_{\{A \text{ and } \tau \leqq t\}} X_\tau\, d\mathscr{P} + \int_{\{A \text{ and } \tau > t\}} X_t\, d\mathscr{P}.\end{aligned}$$

Let $t \to \infty$; use (b) and (c). ★

Theorem (30) partially extends (28). To state it, let

$$\{X_t, \mathscr{F}_t : t = 0, 1, \ldots\}$$

be a nonnegative, expectation-decreasing martingale. Let $\tau_0 \leqq \tau_1 \leqq \cdots$ be stopping times. Let

$$Y_n = X_{\tau_n}.$$

Let $\mathscr{G}_n$ be the σ-field of measurable A such that

$$\{A \text{ and } \tau_n \leqq t\} \in \mathscr{F}_t$$

for all $t = 0, 1, \ldots$.

(30) Theorem. $\{Y_n, \mathscr{G}_n : n = 0, 1, \ldots\}$ *is an expectation-decreasing martingale.*

PROOF. You should check that Y_n is $\mathscr{G}_n$-measurable, and $\mathscr{G}_n \subset \mathscr{G}_{n+1}$. Let $A \in \mathscr{G}_n$. I have to prove

$$\int_A Y_n\, d\mathscr{P} \geqq \int_A Y_{n+1}\, d\mathscr{P}.$$

It is enough to do this with $A \cap \{\tau_n = m\}$ in place of A: afterwards, sum over m. Abbreviate $\sigma = \tau_{n+1}$, so $Y_{n+1} = X_\sigma$. On $A \cap \{\tau_n = m\}$,

$$\sigma \geqq m \quad \text{and} \quad Y_n = X_m.$$

Check $\{\tau_n = m\}$ and $A \cap \{\tau_n = m\}$ are both in $\mathscr{F}_m$. My problem is reduced to showing that

$$\int_A X_m \, d\mathscr{P} \geqq \int_A Y_{n+1} \, d\mathscr{P}, \tag{31}$$

for a typical $A \in \mathscr{F}_m$ with $A \subset \{\tau_n = m\}$. I will argue inductively that for $M \geqq m$,

$$\int_A X_m \, d\mathscr{P} \geqq \int_{A \cap \{\sigma \leqq M\}} X_\sigma \, d\mathscr{P} + \int_{A \cap \{\sigma > M\}} X_m \, d\mathscr{P}. \tag{32}$$

This is clear for $M = m$. Suppose it true for some $M \geqq m$. Now

$$\{\sigma > M\} = \Omega \setminus \{\sigma \leqq M\} \in \mathscr{F}_M, \quad \text{and} \quad A \in \mathscr{F}_m \subset \mathscr{F}_M.$$

The computation rolls on:

$$\begin{aligned}
\int_{A \cap \{\sigma > M\}} X_M \, d\mathscr{P} &\geqq \int_{A \cap \{\sigma > M\}} X_{M+1} \, d\mathscr{P} \\
&= \int_{A \cap \{\sigma = M+1\}} X_{M+1} \, d\mathscr{P} + \int_{A \cap \{\sigma > M+1\}} X_{M+1} \, d\mathscr{P} \\
&= \int_{A \cap \{\sigma = M+1\}} X_\sigma \, d\mathscr{P} + \int_{A \cap \{\sigma > M+1\}} X_{M+1} \, d\mathscr{P}.
\end{aligned}$$

This proves (32). Now (32) is even truer without the rightmost term. Drop it, and let M increase to ∞ to get (31). ★

If $s_1, \ldots, s_N$ is a sequence of real numbers, and $a < b$ are real numbers, the number of *downcrossings* of $[a, b]$ by $s_1, \ldots, s_N$ is the largest positive integer k for which there exists integers $1 \leqq n_1 < n_2 < \cdots < n_{2k} \leqq N$ with

$$s_{n_1} \geqq b, \; s_{n_2} \leqq a, \ldots, s_{n_{2k-1}} \geqq b, \; s_{n_{2k}} \leqq a.$$

If no such k exists, the number of downcrossings is 0.

(33) Downcrossings inequality. *Let $X_0, X_1, \ldots$ be a nonnegative, expectation-decreasing martingale. Let $0 \leqq a < b < \infty$. The mean number of downcrossings of $[a, b]$ by $X_0, X_1, \ldots$ is at most $(b - a)^{-1}$ times the mean of X_0.*

This differs only in detail from the upcrossings inequality (Doob, 1953, Theorem 3.3 on p. 316).

PROOF. Introduce β_n for the number of downcrossings of $[a, b]$ by $X_0, \ldots, X_n$. It is enough to prove

$$E(\beta_n) \leqq E(X_0)/(b - a)$$

in the case $X_i \leqq b$ for $i = 0, \ldots, n$; use (27) on the function $x \to \max\{x, b\}$.

Let σ_0 be the least $m = 0, \ldots, n$ if any with $X_m = b$: if none, let $\sigma_0 = n$. Let σ_1 be the least $m = 0, \ldots, n$ if any with $m > \sigma_0$ and $X_m \leqq a$; if none, let $\sigma_1 = n$. And so on, up to σ_n. Now $X_0, X_{\sigma_0}, \ldots, X_{\sigma_n}$ is an expectation-decreasing martingale by (30). Check

$$\beta_n(b - a) \leqq \Sigma_m \{X_{\sigma_m} - X_{\sigma_{m+1}} : m = 0, \ldots, n - 1 \text{ and } m \text{ is even}\}.$$

Therefore,

$$\begin{aligned} E(X_0) &\geqq E(X_{\sigma_0}) \\ &\geqq E(X_{\sigma_0}) - E(X_{\sigma_n}) \\ &= \Sigma_{m=0}^{n-1} \{E(X_{\sigma_m}) - E(X_{\sigma_{m+1}})\} \\ &\geqq \Sigma_m \{E(X_{\sigma_m}) - E(X_{\sigma_{m+1}}) : m = 0, \ldots, n - 1 \text{ and } m \text{ is even}\} \\ &= E[\Sigma_m \{X_{\sigma_m} - X_{\sigma_{m+1}} : m = 0, \ldots, n - 1 \text{ and } m \text{ is even}\}] \\ &\geqq E[\beta_n(b - a)]. \end{aligned}$$

★

Martingale convergence theorem

(34) Theorem. Forward martingales. *Let $\{X_n : n = 0, 1 \ldots\}$ be a martingale. If* $\sup_n E(|X_n|) < \infty$, *then X_n converges* a.e. *as $n \to \infty$; the convergence is L^1 iff the X_n are uniformly integrable. If $p > 1$ and* $\sup_n E(|X_n|^p) < \infty$, *then X_n converges in L^p as $n \to \infty$. Suppose $\{\mathscr{F}_n : n = 0, 1, \ldots\}$ are nondecreasing σ-fields and $X \in L^p$ for $p \geqq 1$ and $X_n = E\{X|\mathscr{F}_n\}$. Then $\{X_n\}$ is a martingale, and $E(|X_n|^p) \leqq E(|X|^p)$, so the previous assertions apply: The X_n are automatically uniformly integrable, even for $p = 1$. The limit of X_n is the conditional expectation of X given the σ-field generated by all the $\mathscr{F}_n$.*

Backward martingales. *Let $\{X_n : n = \ldots, -3, -2, -1\}$ be a martingale. Then X_n converges* a.e. *as $n \to -\infty$. The X_n are automatically uniformly integrable, and the convergence is also L^1. If $X_{-1} \in L^p$ for $p > 1$, so are all the X_n, and the convergence is also L^p. Suppose $\{\mathscr{F}_n : = \ldots, -3, -2, -1\}$ are nondecreasing σ-fields and $X \in L^p$ for $p \geqq 1$ and $X_n = E\{X|\mathscr{F}_n\}$. Then $\{X_n\}$ is a martingale and $E(|X_n|^p) \leqq E(|X|^p)$, so the previous assertions apply. The limit of X_n is the conditional expectation of X given the intersection of the $\mathscr{F}_n$.*

Proof. If the X_n are nonnegative, the a.e. convergence follows from (33). General X_n follow the same route, with minor changes. The L^p convergence follows from (6). ★

Differentiation

For (35–36), let P and Q be two probabilities on $\mathscr{F}$. Then Ω divides up into three $\mathscr{F}$-sets, Ω_P and Ω_e and Ω_Q, such that $P(\Omega_Q) = Q(\Omega_P) = 0$ and P is equivalent to Q on Ω_e. This partition is unique up to $(P + Q)$-null sets. Let

$$\frac{dP}{dQ} = 0 \quad \text{on } \Omega_Q$$
$$= \infty \quad \text{on } \Omega_P$$

and let dP/dQ be the Radon-Nikodym derivative of P with respect to Q on Ω_e. This function is $\mathscr{F}$-measurable and unique up to changes on $(P + Q)$-null sets. Let $\{\mathscr{A}_n\}$ be a nondecreasing or nonincreasing sequence of σ-fields. In the former case, let $\mathscr{A}_\infty$ be the σ-field generated by the union of the $\mathscr{A}_n$. In the latter case, let $\mathscr{A}_\infty$ be the intersection of the $\mathscr{A}_n$. For any measure R, let R_n be the retraction of R to $\mathscr{A}_n$. Define dP_n/dQ_n like dP/dQ above, with $\mathscr{A}_n$ replacing $\mathscr{F}$. Thus, dP_n/dQ_n is an $\mathscr{A}_n$-measurable function to $[0, \infty]$, unique up to changes on $(P_n + Q_n)$-null sets.

(35) Theorem. $\lim_{n\to\infty} \dfrac{dP_n}{dQ_n} = \dfrac{dP_\infty}{dQ_\infty}$, *except on a $(P + Q)$-null set.*

PROOF. Introduce the probability $A = \frac{1}{2}(P + Q)$. Then $P \leqq 2A$, so $P_n \ll A_n$ and $0 \leqq dP_n/dA_n \leqq 2$. Let E_A be expectation relative to A. Abbreviate $r_n = dP_n/dQ_n$. Suppose the $\mathscr{A}_n$ are nondecreasing. Make the convention $\infty/\infty = 1$, and use (11) to check that

$$\frac{2r_n}{1 + r_n} = \frac{dP_n}{dA_n} = E_A\left\{\frac{dP_\infty}{dA_\infty}\middle|\mathscr{A}_n\right\};$$

and use (34). Suppose the $\mathscr{A}_n$ are nonincreasing. Check that

$$\frac{2r_n}{1 + r_n} = \frac{dP_n}{dA_n} = E_A\left\{\frac{dP_1}{dA_1}\middle|\mathscr{A}_n\right\},$$

and use (34). ★

(36) Example. Suppose the $\mathscr{A}_n$ are nondecreasing. If $P_n \ll Q_n$ for $n = 1, 2, \ldots$, then $\{dP_n/dQ_n : n = 1, 2, \ldots\}$ is a martingale.

9. METRIC SPACES†

A *metric* ρ on Ω is a nonnegative real-valued function on $\Omega \times \Omega$ such that: $\rho(x, y) = 0$ iff $x = y$; and

$$\rho(x, y) + \rho(y, z) \geqq \rho(x, z).$$

† References: (Dunford and Schwartz, 1958), (Hausdorff, 1957), (Kuratowski, 1958), (Loève, 1963, Section 2).

Say x_n *converges to* x or $x_n \to x$ iff $\rho(x_n, x) \to 0$. Say $V \subset \Omega$ is *closed* iff $x_n \in V$ and $x_n \to x$ implies $x \in V$. Say $U \subset \Omega$ is *open* iff $\Omega \setminus U$ is closed. The *topology* or *ρ-topology* of Ω is the set of open U. A sequence $\{x_n\}$ in Ω is *Cauchy* iff $\rho(x_n, x_m) \to 0$ as $n, m \to \infty$. Say Ω is *complete* iff each Cauchy sequence converges. Say Ω is *separable* iff there is a countable subset C of Ω *dense* in Ω: for each $x \in \Omega$, there are $x_n \in C$ with $x_n \to x$.

If (Ω, ρ) is complete and separable, the *Borel σ-field* of (Ω, ρ) is the smallest σ-field containing the ρ-topology. Then $(\Omega_1, \mathscr{F}_1)$ is *Borel* iff there is a complete, separable metric space (Ω, ρ), such that Ω_1 is in the Borel σ-field $\mathscr{F}$ of (Ω, ρ), and $\mathscr{F}_1 = \Omega_1 \mathscr{F}$. In this case also, $\mathscr{F}_1$ is called the *Borel σ-field* of Ω_1. If $(\Omega_i, \mathscr{F}_i)$ are Borel, so is $(\Omega_1 \times \Omega_2 \times \cdots, \mathscr{F}_1 \times \mathscr{F}_2 \times \cdots)$. Borel σ-fields are countably generated.

If I is a countably infinite set, $\rho(i, j) = 1$ or 0 according as $i \neq j$ or $i = j$ is a perfectly good metric; I is complete and separable. The corresponding topology and σ-field are called *discrete*. The *one-point compactification* $\bar{I} = I \cup \{\varphi\}$ is obtained as follows. Let $I = \{i_1, i_2, \ldots\}$; let $\varphi \notin I$; and let $\bar{I} = \{i_1, i_2, \ldots, i_\infty\}$ with $i_\infty = \varphi$; let

$$\bar{\rho}(i_m, i_n) = \left| \frac{1}{m} - \frac{1}{n} \right|,$$

where $1/\infty = 0$. Of course, $\bar{\rho}$ retracted to I produces the same topology as ρ.

(37) Example. Let Ω be Euclidean n-space R^n. Let ρ be the usual distance: $\rho(x, y) = \|x - y\|$ and $\|u\|^2 = \Sigma_{i=1}^n u_i^2$. Then Ω is complete and separable.

(38) Example. Let Ω be the rationals. There is no way to metrize the usual topology so that Ω is complete.

10. REGULAR CONDITIONAL DISTRIBUTIONS†

COMFORT. The material in Sections 10–15 is fairly exotic, and is used only on special occasions.

Let X_i be a measurable mapping from Ω to $(\Omega_i, \mathscr{F}_i)$ for $i = 1, 2$. A *regular conditional $\mathscr{P}$-distribution* for X_2 given X_1 is a function $Q(\cdot, \cdot)$ on $\Omega_1 \times \mathscr{F}_2$ with the following properties:

(39a) $Q(x_1, \cdot)$ is a probability on $\mathscr{F}_2$ for each $x_1 \in \Omega_1$;

(39b) $Q(\cdot, A_2)$ is $\mathscr{F}_1$-measurable for each $A_2 \in \mathscr{F}_2$;

† References: (Blackwell, 1954); (Loève, 1963, Sections 26 and 27).

(39c) for each $A_2 \in \mathscr{F}_2$, the function $\omega \to Q[X_1(\omega), A_2]$ is a version of

$$\mathscr{P}\{X_2 \in A_2 | X_1^{-1}\mathscr{F}_1\}.$$

Condition (c) can be rephrased as follows: if $A_i \in \mathscr{F}_i$, then

(39d) $$\int_{A_1} Q(x_1, A_2)(\mathscr{P}X_1^{-1})(dx_1) = \mathscr{P}\{X_1 \in A_1 \text{ and } X_2 \in A_2\}.$$

You only have to check this for generating classes of A_1's and A_2's closed under intersection, by (16). Sometimes $Q(x_1, \cdot)$ is called a regular conditional $\mathscr{P}$-distribution for X_2 given $X_1 = x_1$.

Suppose Q is a regular conditional $\mathscr{P}$-distribution for X_2 given X_1. Let ϕ be a measurable mapping from $(\Omega_2, \mathscr{F}_2)$ to $(\Omega_\phi, \mathscr{F}_\phi)$. Let $Q_\phi(x_1, \cdot)$ be the $Q(x_1, \cdot)$-distribution of ϕ. Make sure that $Q_\phi(x_1, \cdot)$ is a probability on Ω_ϕ.

(40) Lemma. *Q_ϕ is a regular conditional $\mathscr{P}$-distribution for $\phi(X_2)$ given X_1.*

EXAMPLE. Let $\Omega_1 = \Omega_2 = \Omega = [0, 1]$. Let $X_1(\omega) = X_2(\omega) = \omega$. Let $\mathscr{F}_1$ be the Borel σ-field of $[0, 1]$. Let λ be Lebesgue measure on $\mathscr{F}_1$. Let B be a non-Lebesgue-measurable set, namely $\lambda_*(B) < \lambda^*(B)$. Let $\mathscr{F}_2 = \mathscr{F}$ be the σ-field generated by $\mathscr{F}_1$ and B. Extend λ to a probability $\mathscr{P}$ on $\mathscr{F}$. You can do this so $\mathscr{P}(B)$ is any number in the interval $[\lambda_*(B), \lambda^*(B)]$. There is no regular conditional $\mathscr{P}$-distribution for X_2 given X_1. For suppose Q were such an object. Theorem (51) below produces a $\mathscr{P}$-null set $N \in \mathscr{F}_1$, such that $Q(\omega, \cdot)$ is point mass at ω for $\omega \notin N$. In particular,

$$Q(\omega, B) = 1_B(\omega) \quad \text{for } \omega \notin N.$$

The left side is an $\mathscr{F}_1$-measurable function of ω. So B differs by a subset of the null set N from an $\mathscr{F}_1$-set, a contradiction. ★

(41) Theorem. *If $(\Omega_2, \mathscr{F}_2)$ is Borel, then a regular conditional $\mathscr{P}$-distribution for X_2 given X_1 exists.*

Theorem (41) is hard. One of its virtues (although this does not materially increase the difficulty) is the absence of conditions on $\mathscr{F}$ or $\mathscr{F}_1$.

(42) Theorem. *Suppose $\mathscr{F}_2$ is countably generated. Suppose Q and Q^* are two regular conditional $\mathscr{P}$-distributions for X_2 given X_1. Then*

$$\{x : x \in \Omega_1 \text{ and } Q(x, \cdot) = Q^*(x, \cdot)\}$$

is in $\mathscr{F}_1$ and has $\mathscr{P}X_1^{-1}$-probability 1.

PROOF. Let $\mathscr{A}$ be a countable, generating algebra for $\mathscr{F}_2$. Then

$$\{x : x \in \Omega_1 \text{ and } Q(x, A) = Q^*(x, A)\}$$

is an $\mathscr{F}_1$-set of $\mathscr{P}X_1^{-1}$-probability 1, for each $A \in \mathscr{A}$. The intersection over all $A \in \mathscr{A}$ is the set described in the theorem. ★

The next result generalizes converse Fubini (22). Suppose $\Omega = \Omega_1 \times \Omega_2$ and $\mathscr{F} = \mathscr{F}_1 \times \mathscr{F}_2$. Suppose $\mathscr{P}_1$ is a probability on $\mathscr{F}_1$, and Q satisfies (39a, b). Let

$$X_1(x_1, x_2) = x_1 \quad \text{and} \quad X_2(x_1, x_2) = x_2.$$

(43) Theorem. *There is a unique probability $\mathscr{P}$ on $\mathscr{F}$ satisfying the two conditions:*

(a) $\mathscr{P}X_1^{-1} = \mathscr{P}_1$, *and*

(b) *Q is a regular conditional $\mathscr{P}$-distribution for X_2 given X_1.*

If f is a nonnegative, measurable function on $(\Omega_1 \times \Omega_2, \mathscr{F}_1 \times \mathscr{F}_2)$, then

$$\int_\Omega f\,d\mathscr{P} = \int_{\Omega_1}\int_{\Omega_2} f(x_1, x_2)\,Q(x_1, dx_2)\mathscr{P}_1(dx_1).$$

PROOF. The uniqueness follows from (16). For existence, define

$$\mathscr{P}(A) = \int_{\Omega_1} Q(x_1, A(x_1))\,\mathscr{P}_1(dx_1)$$

for $A \in \mathscr{F}_1 \times \mathscr{F}_2$: as before, $A(x_1)$ is the x_1-section of A, namely the set of x_2 with $(x_1, x_2) \in A$. Check that $\mathscr{P}$ is a probability satisfying (a) and (b). The integration formula now holds with $f = 1_A$; both sides are linear and continuous under increasing passages to the limit. ★

Regular conditional distributions given Σ

In the book, the usual case is: $\Omega_1 = \Omega$ and $\mathscr{F}_1 \subset \mathscr{F}$ and $X_1(\omega) = \omega$. Then, a regular conditional $\mathscr{P}$-distribution for X_2 given X_1 is called a regular conditional $\mathscr{P}$-distribution for X_2 given $\mathscr{F}_1$. The next theorem (43) embodies the main advantage of regular distributions. It is easy to prove, and intuitive: it says that when you condition on a σ-field Σ, you can put any Σ-measurable function U equal to a typical value u, and then substitute U for u when you're through conditioning. That is, U is truly constant given Σ. However, example (48) shows that something a little delicate happened.

I will state (44) in its most popular form. The notation will be used through (50). Let $(\Omega, \mathscr{F}, \mathscr{P})$ be the basic probability triple. Let Σ be a sub-σ-field of $\mathscr{F}$. Let U be a measurable mapping from (Ω, Σ) to a new space $(\Omega_U, \mathscr{F}_U)$. Let V be a measurable mapping from $(\Omega, \mathscr{F})$ to a new space $(\Omega_V, \mathscr{F}_V)$. Thus, U is Σ-measurable and V is $\mathscr{F}$-measurable. The situation is summarized in Figure 1. Let Q be a regular conditional $\mathscr{P}$-distribution for V given Σ, so Q is a function of pairs (ω, C) with $\omega \in \Omega$ and $C \in \mathscr{F}_V$. Let F be a nonnegative, measurable function on $(\Omega_U \times \Omega_V, \mathscr{F}_U \times \mathscr{F}_V)$.

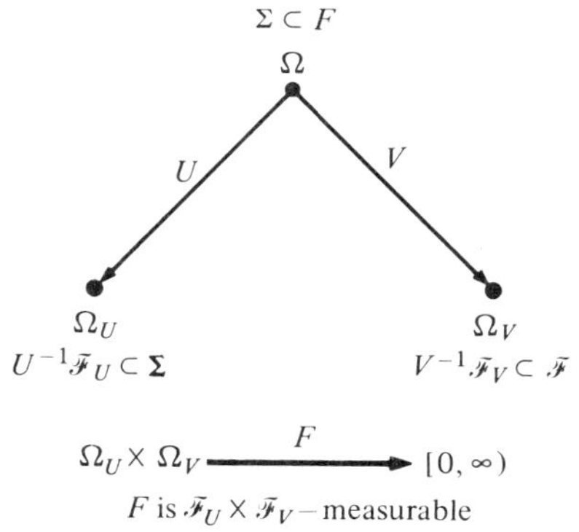

Figure 1.

(44) Theorem. *Let* $F^*(u, \omega) = \int_{\Omega_V} F(u, v)\, Q(\omega, dv)$. *Then* F^* *is* $\mathscr{F}_U \times \Sigma$-*measurable. And* $F^*(U, \cdot)$, *namely*

$$\omega \to F^*(U(\omega), \omega),$$

is a version of $E\{F(U, V)|\Sigma\}$.

PROOF. You can check the measurability of F^* by using (47) below. Now $\omega \to (U(\omega), \omega)$ is a measurable mapping from (Ω, Σ) to $(\Omega_U \times \Omega, \mathscr{F}_U \times \Sigma)$. But $F^*(U, \cdot)$ is the composition of F^* with this mapping, and is Σ-measurable. Fix $A \in \Sigma$. I have to show

$$\int_A \int_{\Omega_V} F(U(\omega), v)\, Q(\omega, dv)\, \mathscr{P}(d\omega) = \int_A F(U(\omega), V(\omega))\mathscr{P}(d\omega). \tag{45}$$

I know

$$\int_S Q(\omega, C)\, \mathscr{P}(d\omega) = \mathscr{P}\{S \text{ and } V \in C\}$$

for $S \in \Sigma$ and $C \in \mathscr{F}_V$. Rewrite this with $\{A \text{ and } U \in B\}$ in place of S, where B is a variable element of $\mathscr{F}_U$. This is legitimate because U is Σ-measurable. I now have (45) for a special F:

$$F(u, v) = 1_B(u) \cdot 1_C(v).$$

Both sides of (45) are linear in F, and continuous under increasing passages to the limit. Use (47) below. ★

(46) Corollary. $E[F(U, V)] = \int_\Omega \int_{\Omega_V} F(U(\omega), v)\, Q(\omega, dv)\, \mathscr{P}(d\omega)$.

(47) Lemma. *Let* F *be a family of nonnegative,* $(\mathscr{F}_U \times \mathscr{F}_V)$-*measurable functions on* $\Omega_U \times \Omega_V$. *Suppose* $af + bg \in$ F *when* $f, g \in$ F *and* a, b *are nonnegative constants. Suppose* $f - g \in$ F *when* $f, g \in$ F *and* $1 \geqq f \geqq g$. *Suppose*

$f \in \mathrm{F}$ when $f_n \in \mathrm{F}$ and $f_n \uparrow f$. Finally, suppose

$$(u, v) \to 1_B(u) \cdot 1_C(v)$$

is in F when $B \in \mathscr{F}_U$ and $C \in F_V$. Then F consists of all the nonnegative measurable functions on $(\Omega_U \times \Omega_V, \mathscr{F}_U \times \mathscr{F}_V)$.

(48) Example. Suppose $U = V$ is uniform on $[0, 1]$. Let $F(u, v)$ be 1 or 0 according as $u = v$ or $u \neq v$. Then $F(U, V) = 1$ almost surely, so

$$E\{F(U, V)|U\} = 1$$

almost surely. But $F(u, V) = 0$ almost surely for any particular u, forcing $E\{F(u, V)|U\} = 0$ almost surely. Theorem (43) rescues this example by defining

$$E\{F(u, V)|U = u\} = 1. \qquad \bigstar$$

Theorem (49) sharpens (44). To state it and (50), let ϕ be a measurable mapping from $(\Omega_U \times \Omega_V, \mathscr{F}_V \times \mathscr{F}_V)$ to some new space $(\Omega_\phi, \mathscr{F}_\phi)$. Temporarily, fix $\omega \in \Omega$ and $u \in \Omega_U$. Then $\phi(u, \cdot)$ is a measurable mapping from $(\Omega_V, \mathscr{F}_V)$ to $(\Omega_\phi, \mathscr{F}_\phi)$. And $Q(\omega, \cdot)$ is a probability on $\mathscr{F}_V$. So I am entitled to define $D(\omega, u, \cdot)$ as the $Q(\omega, \cdot)$-distribution of $\phi(u, \cdot)$; this comes out to a probability on $\mathscr{F}_\phi$. Let $R(\omega, \cdot) = D(\omega, U(\omega), \cdot)$, again a probability on $\mathscr{F}_\phi$.

(49) Theorem. *$R(\cdot, \cdot)$ is a regular conditional $\mathscr{P}$-distribution for $\phi(U, V)$ given Σ.*

PROOF. Let $A \in \Sigma$ and $B \in \mathscr{F}_\phi$. I have to check that

$$\mathscr{P}\{A \text{ and } \phi(U, V) \in B\} = \int_A R(\omega, B)\, \mathscr{P}(d\omega).$$

Let $F(u, v) = 1$ or 0, according as $\phi(u, v) \in B$ or $\notin B$. Use (44):

$$\begin{aligned}\mathscr{P}\{A \text{ and } \phi(U, V) \in B\} &= \int_A F(U, V)\, d\mathscr{P} \\ &= \int_A E\{F(U, V)|\Sigma\}\, d\mathscr{P} \\ &= \int_A \int_{\Omega_V} F[U(\omega), v]\, Q(\omega, dv)\, \mathscr{P}(d\omega).\end{aligned}$$

Recognize

$$\begin{aligned}\int_{\Omega_V} F[U(\omega), v]\, Q(\omega, dv) &= Q(\omega, \{v : \phi[U(\omega), v] \in B\}) \\ &= R(\omega, B). \qquad \bigstar\end{aligned}$$

The situation is more tractable when Σ and V are independent, which will be assumed in (50). Let $D(u, \cdot)$ be the $\mathscr{P}$-distribution of $\phi(u, \cdot)$, a probability on $\mathscr{F}_\phi$ for each $u \in \Omega_U$.

(50) Theorem. *Suppose Σ and V are independent. Then $D(U, \cdot)$ is a regular conditional $\mathscr{P}$-distribution for $\phi(U, V)$ given Σ.*

PROOF. Let $A \in \Sigma$ and $B \in \mathscr{F}_\phi$. I have to check that

$$\mathscr{P}\{A \text{ and } \phi(U, V) \in B\} = \int_A D[U(\omega), B]\,\mathscr{P}(d\omega).$$

Use Fubini (21) to evaluate the left side. Keep $(\Omega, \mathscr{F}, \mathscr{P})$ for the basic probability triple. Put (Ω, Σ) for $(\Omega_1, \mathscr{F}_1)$, with $X_1(\omega) = \omega$. Put $(\Omega_V, \mathscr{F}_V)$ for $(\Omega_2, \mathscr{F}_2)$, with $X_2(\omega) = V(\omega)$. Let $f(\omega, v) = 1$ if $\omega \in A$ and $\phi[U(\omega), v] \in B$; otherwise, let $f(\omega, v) = 0$. Let $\hat{\mathscr{P}}$ be $\mathscr{P}$ retracted to Σ. Then

$$\begin{aligned}\mathscr{P}\{A \text{ and } \phi(U, V) \in B\} &= \int_\Omega f[X_1(\omega), V(\omega)]\,\mathscr{P}(d\omega)\\ &= \int_\Omega \int_\Omega f[\omega, V(\omega')]\,\mathscr{P}(d\omega')\,\hat{\mathscr{P}}(d\omega)\\ &= \int_A \int_\Omega f[\omega, V(\omega')]\,\mathscr{P}(d\omega')\,\hat{\mathscr{P}}(d\omega),\end{aligned}$$

because $f(\omega, \cdot) \equiv 0$ for $\omega \notin A$. Recognize

$$\begin{aligned}\int_\Omega f[\omega, V(\omega')]\,\mathscr{P}(d\omega') &= \mathscr{P}\{\omega' : \phi[U(\omega), V(\omega')] \in B\}\\ &= D[U(\omega), B]\end{aligned}$$

for $\omega \in A$. ★

Regular conditional probabilities

If $\Omega_V = \Omega$ and $\mathscr{F}_V = \mathscr{F}$ and $V(\omega) = \omega$, then a regular conditional $\mathscr{P}$-distribution for V given Σ is called a regular conditional $\mathscr{P}$-probability given Σ. For (51) and (52), let Q be a regular conditional $\mathscr{P}$-probability given Σ. That is, $(\Omega, \mathscr{F}, \mathscr{P})$ is the basic probability triple, and Σ is a sub-σ-field of $\mathscr{F}$. Moreover Q is a function of pairs (ω, B), with $\omega \in \Omega$ and $B \in \mathscr{F}$. The function $Q(\cdot, B)$ is a version of $\mathscr{P}(B|\Sigma)$, and the function $Q(\omega, \cdot)$ is a probability. Recall that $\Sigma(\omega)$ is the Σ-atom containing ω.

(51) Theorem. *Let Σ be countably generated. Then the set of ω such that $Q(\omega, \Sigma(\omega)) = 1$ is a Σ-set of $\mathscr{P}$-probability* 1.

PROOF. Let $\mathscr{A}$ be a countable generating algebra for Σ. For each $A \in \mathscr{A}$, let A^* be the set of ω such that $Q(\omega, A) = 1_A(\omega)$. Then A^* is a Σ-set of $\mathscr{P}$-probability 1, and the intersection of A^* as A varies over $\mathscr{A}$ is the set described in the theorem. ★

For (52), do not assume that Σ is countably generated. Let $\mathscr{E}$ be the smallest σ-field over which $\omega \to Q(\omega, A)$ is measurable, for all $A \in \mathscr{F}$. Thus, $\mathscr{E} \subset \Sigma$.

Let E be the set of ω such that

$$Q(\omega, \{\omega' : Q(\omega', \cdot) = Q(\omega, \cdot)\}) = 1.$$

(52) Theorem. *Suppose $\mathscr{F}$ is countably generated.*

(a) *$\mathscr{E}$ is countably generated.*

(b) $E \in \mathscr{E}$.

(c) $\mathscr{P}(E) = 1$.

PROOF. Let $\mathscr{A}$ be a countable generating algebra for $\mathscr{F}$. Then $\mathscr{E}$ is also the smallest σ-field over which $\omega \to Q(\omega, A)$ is measurable, for all $A \in \mathscr{A}$, by the monotone class argument (Section 5). This proves (a). As (18) now implies,

$$\mathscr{E}(\omega) = \{\omega' : Q(\omega', A) = Q(\omega, A) \text{ for all } A \in \mathscr{A}\}.$$

Of course, Q is a regular conditional $\mathscr{P}$-probability given $\mathscr{E}$. Finally, (51) proves (b) and (c). ★

Regular conditional distributions for partially defined random variables

Let $(\Omega, \mathscr{F}, \mathscr{P})$ be the basic probability triple, and let Σ be a sub-σ-field of $\mathscr{F}$. Let $D \in \Sigma$. Let V be a measurable mapping from $(D, D\Sigma)$ to a new space $(\Omega_V, \mathscr{F}_V)$. As usual, $D\Sigma$ is the σ-field of subsets of D of the form $D \cap S$ with $S \in \Sigma$. A *regular conditional $\mathscr{P}$-distribution for V given Σ on D* is a function Q of pairs (ω, B) with $\omega \in D$ and $B \in \mathscr{F}_V$, such that:

$Q(\omega, \cdot)$ is a probability on $\mathscr{F}_V$ for each $\omega \in D$;

$Q(\cdot, B)$ is $D\Sigma$-measurable for each $B \in \mathscr{F}_V$; and

$$\int_A Q(\omega, B)\, \mathscr{P}(d\omega) = \mathscr{P}\{A \text{ and } V \in B\}$$

for all $A \in \Sigma$ with $A \subset D$ and all $B \in \mathscr{F}_V$. Of course, A and B can be confined to generating classes in the sense of (16). The partially defined situation is isomorphic to a fully defined one. Replace Ω by D, and $\mathscr{F}$ by $D\mathscr{F}$, and Σ by $D\Sigma$, and $\mathscr{P}$ by $\mathscr{P}\{\cdot \mid D\}$. Theorems like (44) can therefore be used in partially defined situations.

Conditional independence

Let $(\Omega, \mathscr{F})$ and $(\Omega_i, \mathscr{F}_i)$ be Borel. Let $\mathscr{P}$ be a probability on $\mathscr{F}$, and X_i a measurable mapping from $(\Omega, \mathscr{F})$ to $(\Omega_i, \mathscr{F}_i)$. Let Σ be a sub-σ-field of $\mathscr{F}$. What does it mean to say X_1 and X_2 are *conditionally $\mathscr{P}$-independent given*

Σ? The easiest criterion is

$$\mathscr{P}\{X_1 \in A_1 \text{ and } X_2 \in A_2 | \Sigma\} = \mathscr{P}\{X_1 \in A_1 | \Sigma\} \cdot \mathscr{P}\{X_2 \in A_2 | \Sigma\} \quad \text{a.e.}$$

for all $A_i \in \mathscr{F}_i$. Nothing is changed if A_i is confined to a generating class for $\mathscr{F}_i$ in the sense of (16). Here is an equivalent criterion. Let $Q(\cdot, \cdot)$ be a regular conditional $\mathscr{P}$-distribution for (X_1, X_2) given Σ. Then $Q(\omega, \cdot)$ is a probability on $\mathscr{F}_1 \times \mathscr{F}_2$. The variables X_1, X_2 are conditionally $\mathscr{P}$-independent given Σ iff for $\mathscr{P}$-almost all ω,

$$Q(\omega, \cdot) = Q_1(\omega, \cdot) \times Q_2(\omega, \cdot)$$

where $Q_i(\omega, \cdot)$ is the projection of $Q(\omega, \cdot)$ onto $\mathscr{F}_i$. Necessarily, Q_i is a regular conditional $\mathscr{P}$-distribution for X_i given Σ. The equivalence of these conditions is easy, using (10a).

11. THE KOLMOGOROV CONSISTENCY THEOREM†

Let $(\Omega_i, \mathscr{F}_i)$ be Borel for $i = 1, 2, \ldots$. Let

$$\Omega = \Omega_1 \times \Omega_2 \times \cdots \quad \text{and} \quad \mathscr{F} = \mathscr{F}_1 \times \mathscr{F}_2 \times \cdots.$$

Let π_n *project* $\Omega_1 \times \cdots \times \Omega_{n+1}$ onto $\Omega_1 \times \cdots \times \Omega_n$: namely,

$$\pi_n(\omega_1, \ldots, \omega_n, \omega_{n+1}) = (\omega_1, \ldots, \omega_n).$$

Let $\boldsymbol{\pi}_n$ *project* Ω onto $\Omega_1 \times \cdots \times \Omega_n$: namely,

$$\boldsymbol{\pi}_n(\omega_1, \ldots, \omega_n, \omega_{n+1}, \ldots) = (\omega_1, \ldots, \omega_n).$$

For $n = 1, 2, \ldots$, let $\mathscr{P}_n$ be a probability on $(\Omega_1 \times \cdots \times \Omega_n, \mathscr{F}_1 \times \cdots \times \mathscr{F}_n)$. Suppose the $\mathscr{P}_n$ are *consistent*, namely, $\mathscr{P}_{n+1}\pi_n^{-1} = \mathscr{P}_n$ for all n.

(53) Theorem. *There is a unique probability $\mathscr{P}$ on $(\Omega, \mathscr{F})$ with $\mathscr{P}\boldsymbol{\pi}_n^{-1} = \mathscr{P}_n$ for all n.*

12. THE DIAGONAL ARGUMENT

Let Z be the positive integers. Let S be the set of strictly increasing functions from Z to Z. Call $s \in S$ a *subsequence* of Z. For $s \in S$, the *range* of s is the s-image $s(Z)$ of Z; and $s(n) \geqq n$. Say s is a *subsequence* or on special occasions a *sub-subsequence* of $t \in S$ iff $s \in S$ and for each $n \in Z$, there is a $\sigma(n) \in Z$ with $s(n) = t[\sigma(n)]$. This well-defines σ, and forces $\sigma \in S$. Further, $s(n) \geqq t(n)$, because $\sigma(n) \geqq n$. Geometrically, $s \in S$ is a subsequence of $t \in S$ iff the range

† References: (Loève, 1963, Section 4.3); (Neveu, 1965, Section III.3).

of s is a subset of the range of t. Thus, if s is a subsequence of t, and t is a subsequence of $u \in S$, then s is a subsequence of u. If $s \in S$, and $m = 0, 1, \dots$, define $s(m + \cdot) \in S$ as follows:

$$s(m + \cdot)(n) = s(m + n) \quad \text{for } n \in Z.$$

Of course, $s(m + \cdot)$ is a subsequence of s.

Here is a related notion. Say s is *eventually* a subsequence of $t \in S$ iff $s \in S$ and $s(m + \cdot)$ is a subsequence of t for some $m = 0, 1, \dots$. Geometrically, $s \in S$ is eventually a subsequence of $t \in S$ iff the range of s differs by a finite set from a subset of the range of t. In particular, if s is eventually a subsequence of t, and t is eventually a subsequence of $u \in S$, then s is eventually a subsequence of u.

To state the *first diagonal principle*, let $s_1 \in S$ and let s_{n+1} be a subsequence of s_n for $n = 1, 2, \dots$. Let d be the *diagonal sequence*:

$$d(n) = s_n(n) \quad \text{for } n = 1, 2, \dots.$$

(54) First diagonal principle. *The diagonal sequence d is a subsequence of s_1, and is eventually a subsequence of s_n for all n.*

PROOF. I claim $d \in S$:

$$d(n + 1) = s_{n+1}(n + 1) \geqq s_n(n + 1) > s_n(n) = d(n).$$

Fix $n = 1, 2, \dots$. I claim $d(n - 1 + \cdot)$ is a subsequence of s_n. Indeed, fix $m = 1, 2, \dots$. Then $m - 1 \geqq 0$, so

$$d(n - 1 + m) = s_{n-1+m}(n - 1 + m) \in s_{n-1+m}(Z) \subset s_n(Z). \qquad \bigstar$$

(55) Illustration. Let $s_n(m) = n + m$. So $d(n) = 2n$, as in Figure 2.

To make this a little more interesting, introduce a metric space (Ω, ρ). Let f be a function from Z to Ω. Let $t \in S$. Suppose

$$\lim_{n\to\infty} f[t(n)] = y \in \Omega.$$

If s is eventually a subsequence of t, you should check

$$\lim_{n\to\infty} f[s(n)] = y.$$

For the *second diagonal principle*, let C be a countable set. For $c \in C$, let f_c be a function from Z to Ω. Suppose that for each $c \in C$ and $t \in S$, there is a subsequence s_c of t, such that

$$\lim_{n\to\infty} f_c[s_c(n)]$$

exists. This s_c depends on c and t.

$s \backslash m$	1	2	3	4	·	·	·	·	·
s_1	2	3	4	5	·	·	·	·	·
s_2	3	4	5	6	·	·	·	·	·
s_3	4	5	6	7	·	·	·	·	·
s_4	5	6	7	8	·	·	·	·	·
·	·	·	·	·	·	·	·	·	·
·	·	·	·	·	·	·	·	·	·
·	·	·	·	·	·	·			
·	·	·	·	·	·	·			
·	·	·	·	·	·	·			

Figure 2.

(56) Second diagonal principle. *For each $t \in S$, there is a subsequence d of t such that*

$$\lim_{n \to \infty} f_c[d(n)]$$

exists for all $c \in C$. This d depends on t, but not on c.

PROOF. Enumerate C as $\{c_1, c_2, \ldots\}$. Abbreviate $f_n = f_{c_n}$. Inductively, construct $s_n \in S$ such that $s_0 = t$ and s_n is a subsequence of s_{n-1} and

$$\lim_{m \to \infty} f_n[s_n(m)]$$

exists. Call this limit y_n; of course, y_n may depend on $s_1, \ldots, s_n$. Using the first diagonal principle, construct the diagonal subsequence d, which is a subsequence of t and eventually a subsequence of each s_n. So,

$$\lim_{m \to \infty} f_n[d(m)] = y_n. \qquad \bigstar$$

13. CLASSICAL LEBESGUE MEASURE

Euclidean n-space R^n comes equipped with a metric ρ_n, and is complete and separable (37). A real-valued random variable on $(\Omega, \mathscr{F})$ is now a measurable mapping to Borel R^1. The classical *n-dimensional Lebesgue measure* λ_n is the countably additive, nonnegative set function on Borel R^n, whose value at an n-dimensional cube is its n-dimensional volume.

(57) Theorem. *Suppose f is a bounded function on R^n which vanishes outside a cube. Then f is Riemann integrable iff f is continuous λ_n-almost everywhere, and its Riemann integral coincides with $\int f\,d\lambda_n$.*

This theorem will be used only to evaluate Lebesgue integrals, and the Riemann integrability of f will be obvious.

(58) Theorem. *Let f be a measurable function on R^n, with finite Lebesgue integral. Then*

$$\lim_{h\to 0} \int_{R^n} |f(x + h) - f(x)|\lambda_n(dx) = 0.$$

PROOF. If f is continuous and vanishes outside a large cube, the result is clear. General f can be approximated by these special f in L^1-norm. ★

If A is a Borel subset of R^1, the *metric density of A at x is*

$$\lim_{\varepsilon,\delta \downarrow 0} \frac{\lambda_1\{(x - \varepsilon, x + \delta) \cap A\}}{\varepsilon + \delta}.$$

(59) Metric density theorem.[1] *Let A be a Borel subset of R^1. There is a Borel set B with $\lambda_1(B) = 0$, such that A has metric density 1 at all $x \in A \setminus B$.*

14. REAL VARIABLES

Let f be a real-valued function on [0, 1]. Let $S = \{s_0, s_1, \ldots, s_n\}$ be a finite subset of [0, 1] with $0 = s_0 < s_1 < \cdots < s_n = 1$. Let

$$\delta S = \max\{(s_{j+1} - s\ ):j = 0, \ldots, n - 1\},$$

and

$$Sf = \Sigma_{j=0}^{n-1} |f(s_{j+1}) - f(s_j)|.$$

Let $W(S, f) = \Sigma_{j=0}^{n-1} (M_j - m_j)$, where

$$M_j = \max\{f(t):s_j \leqq t \leqq s_{j+1}\} \quad \text{and} \quad m_j = \min\{f(t):s_j \leqq t \leqq s_{j+1}\}.$$

The *variation* of f is $\sup_S Sf$; if this number is finite, f is of *bounded variation.* If S_n is nondecreasing and $\delta S_n \downarrow 0$, then $S_n f$ tends to the variation of f; so $W(S_n, f)$ must tend to the variation of f also.

[1] Reference: (Saks, 1964, Theorem 6.1 on p. 117. Theorem 10.2 on p. 129 is the n-dimensional generalization, which is harder.)

(60) Lebesgue's theorem.[1] *If f is of bounded variation, then f has a finite derivative Lebesgue almost everywhere.*

Theorem (60) can be sharpened as follows.

(61) Theorem.[2] *Suppose f is of bounded variation. The pointwise derivative of f is a version of the Radon–Nikodym derivative of the absolutely continuous part of f, with respect to Lebesgue measure.*

Even more is true.

(62) de la Vallée Poussin's theorem.[3] *Suppose f is of bounded variation. The positive, continuous, singular part of f is concentrated on $\{x:f'(x) = \infty\}$.*

ASSUMPTION. For the rest of this section, assume f is a continuous function on $[0, 1]$.

Let $s(y)$ be the number of x with $f(x) = y$, so $s(y) = 0, 1, \ldots, \infty$.

(63) Banach's theorem.[4] *The variation of f is $\int_{-\infty}^{\infty} s(y)\,dy$.*

PROOF. Let $S_n = \{0, 1/2^n, 2/2^n, \ldots, 1\}$. Let $s_{n,0}$ be the indicator function of the f-image of the interval $[0, 1/2^n]$. For $j = 1, \ldots, 2^n - 1$, let $s_{n,j}$ be the indicator function of the f-image of $(j/2^n, (j + 1)/2^n]$. Let $s_n = \Sigma_{j=0}^{2^n-1} s_{n,j}$. Verify that $s_n \uparrow s$, so s is Borel and

$$\int_{-\infty}^{\infty} s(y)\,dy = \lim_n \int_{-\infty}^{\infty} s_n(y)\,dy = \lim_n W(S_n, f):$$

because

$$\int_{-\infty}^{\infty} s_{n,j}(y)\,dy = M_{n,j} - m_{n,j},$$

where

$$M_{n,j} = \max\{f(t):j/2^n \leqq t \leqq (j + 1)/2^n\}$$
$$m_{n,j} = \min\{f(t):j/2^n \leqq t \leqq (j + 1)/2^n\}.$$ ★

The *upper right Dini derivative* D^*f is defined by:

$$D^*f{\cdot}x = \limsup_{\varepsilon \downarrow 0} [f(x + \varepsilon) - f(x)]/\varepsilon,$$

for $0 \leqq x < 1$.

[1] References: (Saks, 1964, Theorem 5.4 on p. 115); (Riesz-Nagy, 1955, Chapter 1) has a proof from first principles. This theorem is hard.
[2] References: (Dunford and Schwartz, 1958, III.12); (Saks, 1964, Theorem 7.4 on p. 119). It's hard.
[3] Reference: (Saks, 1964, Theorem 9.6 on p. 127). Hard.
[4] Reference: (Saks, 1964, Theorem 6.4 on p. 280).

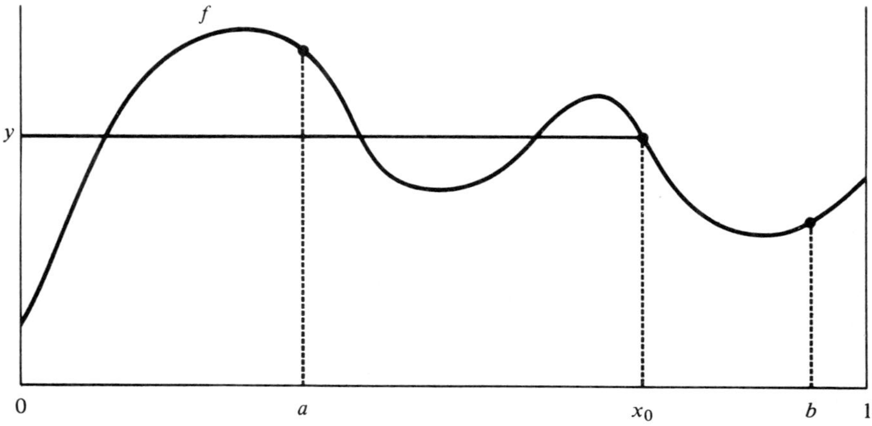

Figure 3.

(64) Zygmund's theorem.[1] *If the set of values assumed by f on the set of x with $D_* f \cdot x \leqq 0$ includes no proper interval, then f is nondecreasing.*

PROOF. Suppose by way of contradiction that there are a and b with $0 < a < b < 1$ and $f(a) > f(b)$. Find one y with $f(a) > y > f(b)$, such that $y \notin f\{D_* f \leqq 0\}$; that is, $y = f(x)$ entails $D_* f \cdot x > 0$. Let x_0 be the largest $x \in [a, b]$ with $f(x) = y$, so $x_0 < b$. But $f < y$ on $(x_0, b]$, so $D_* f \cdot x_0 \leqq 0$. See Figure 3. ★

(65) Corollary. *If the set of x with $D_* f \cdot x < 0$ is at most countable, then f is nondecreasing.*

PROOF. Let $\varepsilon > 0$ and $f_\varepsilon(x) = f(x) + \varepsilon x$. Now $D_* f_\varepsilon \cdot x = D_* f \cdot x + \varepsilon$, so $\{D_* f_\varepsilon \leqq 0\}$ is at most countable. By Zygmund's theorem, f_ε is nondecreasing Let $\varepsilon \to 0$. ★

(66) Dini's theorem.[2]

(a) $$\sup_{0 \leqq x < 1} D^* f \cdot x = \sup_{0 \leqq x < y \leqq 1} \frac{f(y) - f(x)}{y - x}$$

and

(b) $$\inf_{0 \leqq x < 1} D_* f \cdot x = \inf_{0 \leqq x < y \leqq 1} \frac{f(y) - f(x)}{y - x}.$$

[1] Reference: (Saks, 1964, Theorem 7.1 on p. 203).
[2] Reference: (Saks, 1964, p. 204).

PROOF. Fix a finite m. Then $D*f \cdot x \geqq m$ for $0 \leqq x < 1$ implies $f(x) - mx$ is nondecreasing with x, by (65). By algebra, for $0 \leqq x < y \leqq 1$,

$$\frac{f(y) - f(x)}{y - x} \geqq m.$$

Consequently, $\inf[f(y) - f(x)]/(y - x) \geqq \inf$. The opposite inequality is clear, proving (b). Assertion (a) is easy.

(67) Corollary.[1] *If $D*f$ is continuous at x, then f is differentiable at x.*

PROOF. Use (66). ★

(68) Corollary. *If $D*f \equiv 0$ on $[0, 1)$, then f is constant on $[0, 1]$.*

PROOF. Use (67). ★

(69) Theorem. *Suppose f has a finite, right continuous, right derivative f^+ on $(0, 1)$, which has a finite integral over $(\varepsilon, 1 - \varepsilon)$ for any $\varepsilon > 0$. If $0 < x < y < 1$, then*

$$f(y) = f(x) + \int_x^y f^+(t)\, dt.$$

PROOF. Let

$$g(y) = f(y) - f(x) - \int_x^y f^+(t)\, dt.$$

Then g is continuous and $D*g = 0$ on $[x, 1)$, while $g(x) = 0$. Use (68). ★

Miscellany

Let μ be a probability on the Borel subsets of $[0, \infty)$. Its *Laplace transform* φ is this function of nonnegative λ:

$$\varphi(\lambda) = \int_{[0,\infty)} e^{-\lambda x}\, \mu(dx).$$

(70) Theorem.[2] *φ determines μ.*

As usual, $0! = 1$, and $1! = 1$, and $(n + 1)! = (n + 1)n!$. Write

$$\binom{n}{m} = \frac{n!}{m!(n - m)!},$$

the number of subsets with m elements which can be chosen from a set with

[1] Reference: (Saks, 1964, p. 204).
[2] Reference: (Feller, 1966, Theorem 1 on p. 408).

n elements. Temporarily, let

$$s(n) = \sqrt{2\pi n}\left(\frac{n}{e}\right)^n.$$

(71) Stirling's formula.[1] $n!/s(n) \to 1$ as $n \to \infty$.

15. ABSOLUTE CONTINUITY

A function f on $[0, 1]$ is *absolutely continuous* iff for every positive ε there is a positive δ such that:

$$0 \leqq t_1 < t_2 < t_3 < \cdots < t_{2n-1} < t_{2n} \leqq 1$$

and

$$\Sigma_{m=1}^n (t_{2m} - t_{2m-1}) < \delta$$

imply

$$\Sigma_{m=1}^n |f(t_{2m}) - f(t_{2m-1})| < \varepsilon.$$

In other words, f is absolutely continuous iff f is of bounded variation, and the measure induced by f is absolutely continuous with respect to Lebesgue measure. The pointwise derivative of f is then a version of the Radon–Nikodym derivative by (61). Suppose g is another function on $[0, 1]$. If f and g are absolutely continuous, and $f' = g'$ a.e. it follows that $f - g$ is constant. Suppose f and g are absolutely continuous, and g is nondecreasing, and $0 \leqq g(0) \leqq g(1) \leqq 1$. Then $f \circ g$ is absolutely continuous, as is immediate from the definition; and

THE CHAIN RULE.[2] $(f \circ g)' = (f' \circ g)g'$ a.e.

PROOF. For simplicity, suppose $f(0) = g(0) = 0$ and $g(1) = 1$. Let $0 \leqq t \leqq 1$. Here is a computation.

$$\begin{aligned} f \circ g(t) &= \int_0^{g(t)} f'(u)\,du \\ &= \int_0^t f'[g(t)]\,g(dt) \\ &= \int_0^t f'[g(t)]g'(t)\,dt. \end{aligned}$$

[1] Reference: (Feller, 1968, II.9).
[2] Reference: (Serrin and Varberg, 1969).

The first line holds by (61); the second by (20), for the g-distribution of g is uniform on $[0, 1]$; and the third by (61) and (11). Now use (61) and (11) to differentiate. ★

(72) Theorem. *Suppose the upper or lower right Dini derivative of f is finite at all but a countable number of points, and f is of bounded variation. Then f is absolutely continuous.*

PROOF. Use (62). ★

16. CONVEX FUNCTIONS

Let f be a real-valued function on the open interval (a, b). Abbreviate $\bar{p} = 1 - p$. Then f is *convex* iff

$$a < x < y < b \quad \text{and} \quad 0 < p < 1$$

imply

$$f(px + \bar{p}y) \leqq pf(x) + \bar{p}f(y).$$

Geometrically, each chord of f is nowhere below f, as in Figure 4. Say f is *strictly convex* iff inequality holds. Say f is *concave* or *strictly concave* iff $-f$ is convex or strictly convex. For (73), suppose f is convex and

$$a < x < y < b \quad \text{and} \quad a < x' < y' < b$$

$$x \leqq x' \quad \text{and} \quad y \leqq y'.$$

As in Figure 4,

(73) $$\frac{f(y') - f(x')}{y' - x'} \geqq \frac{f(y) - f(x)}{y - x}.$$

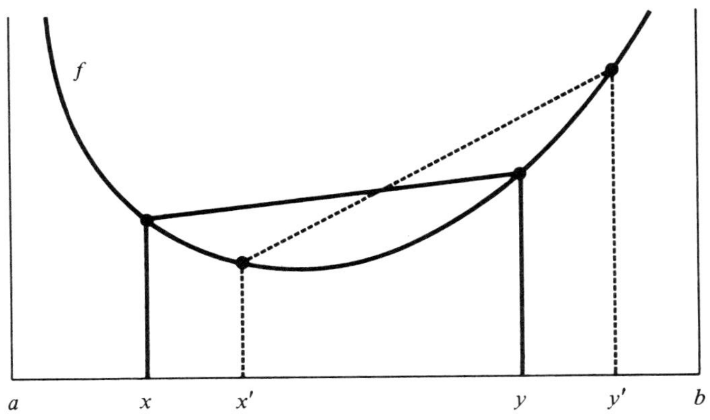

Figure 4.

Indeed, the case $x = x'$ restates the definition of convexity, as does the case $y = y'$. General (73) follows by combining the two cases: the slope of f over (x, y) is at most the slope over (x, y'), which does not exceed the slope over (x', y').

(74) Theorem. *Suppose f is convex on (a, b).*

(a) *f is continuous.*

(b) *f has a finite right derivative f^+, which is nondecreasing and continuous from the right.*

(c) *f has a finite left derivative f^-, which is nondecreasing and continuous from the left.*

(d) $f^+ \geqq f^-$.

(e) *The discontinuity sets of f^+ and f^- coincide, and are countable. Off this set, $f^+ = f^-$.*

(f) *For $a < x < y < b$,*

$$f(y) - f(x) = \int_x^y f^+(t)\,dt = \int_x^y f^-(t)\,dt.$$

(g) *If f is strictly convex, then f^+ and f^- are strictly increasing.*

(h) *Suppose a and $f(a+)$ are finite. Define $f(a) = f(a+)$. Then f has a right derivative $f^+(a)$ at a, and $-\infty \leqq f^+(a) < \infty$. And $f^+(a+) = f^+(a)$. The situation at b is symmetric.*

PROOF. *Claim* (*b*). Let y decrease to $x > w$. The slope of f over (x, y) nonincreases and is at least the slope over (w, x) by (73), proving that

(75) $f^+(x)$ exists and is at most the slope of f over (x, y), and at least the slope over (w, x).

You can use (73) to show that f^+ is nondecreasing. I will argue that f^+ is continuous from the right at x. Let

$$a < x < y < z < b,$$

as in Figure 5.

Then $f^+(x) \leqq f^+(y)$. But $f^+(y)$ is at most the slope of f over (y, z) by (75), which tends to the slope of f over (x, z) when y tends to x. That is,

$$\lim_{y \to x} f^+(y) \leqq \frac{f(z) - f(x)}{z - x} \downarrow f^+(x) \quad \text{as } z \downarrow x.$$

Claim (*c*) is symmetric.

Claim (*d*) follows from (73).

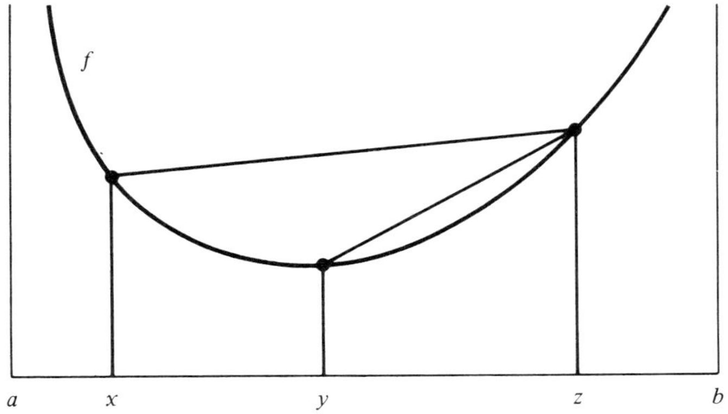

Figure 5.

Claim (a). Get $f(x) = f(x+)$ from the finitude of f^+, and $f(x) = f(x-)$ from symmetry.

Claim (e). Let $h > 0$. Then $f^-(x + h) \geqq f^+(x)$ by (73). And (d) makes $f^-(x + h) \leqq f^+(x + h)$. Let $h \to 0$ and use (b):

$$f^-(x+) = f^+(x+) = f^+(x).$$

Similarly

$$f^-(x-) = f^+(x-) = f^-(x).$$

Claim (f). Use (a, b) and (69), then (e).

Claim (g). Improve (73) to strict inequality, when $x < x'$ or $y < y'$. Let

$$a < x < y < z < w < b.$$

Use (75) and improved (73):

$$f(x+) \leqq \frac{f(y) - f(x)}{y - x} < \frac{f(z) - f(y)}{z - y} < \frac{f(w) - f(z)}{w - z} \downarrow f^+(z)$$

as $w \downarrow z$.

Claim (h) is like (b). ★

(76) Theorem. *Suppose f is convex on (a, b), and $x \in (a, b)$. Using (74d), choose s with*

$$f^-(x) \leqq s \leqq f^+(x).$$

Let

$$\lambda(t) = s(t - x) + f(x),$$

the linear function with slope s which agrees with f at x. Then

$$\lambda \leqq f \quad \text{on } (a, b).$$

In particular, a convex function on a finite interval is bounded below.

PROOF. Let $x < t < b$: the other case is symmetric. Then

$$s \leqq f^+(x) \leqq \frac{f(t) - f(x)}{t - x}$$

by (75); look at Figure 6. So

$$f(t) \geqq s(t - x) + f(x) = \lambda(t). \qquad \bigstar$$

NOTE. Suppose f is a function on (a, b). If either

(a) f has a finite, right continuous derivative f^+ which is nondecreasing, or

(b) there is a nondecreasing g with

$$f(y) = f(x) + \int_x^y g(t)\,dt \quad \text{when } a < x < y < b,$$

then f is convex.

PROOF. Condition (a) implies (b) with $g = f^+$, by (69). Suppose (b). Fix x, y, z with

$$a < x < y < z < b.$$

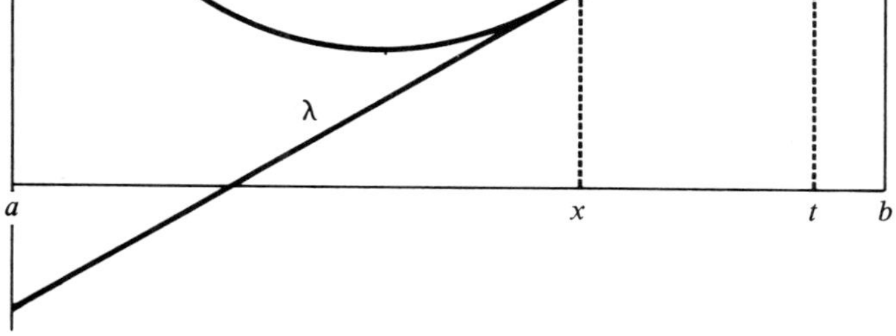

Figure 6.

Abbreviate c for the slope of f over (x, y), and d for the slope of f over (y, z). Then

$$c = \frac{1}{y - x} \int_x^y g(t)\, dt \leqq \frac{1}{z - y} \int_y^z g(t)\, dt = d.$$

So $(y, f(y))$ cannot be above the chord joining $(x, f(x))$ and $(z, f(z))$. ★

17. COMPLEX VARIABLES

If $z = x + iy$, where x and y are real and $i^2 = -1$, then $\operatorname{Re} z = x$ and $\operatorname{Im} z = y$ and $|z| = (x^2 + y^2)^{\frac{1}{2}}$. Moreover, $e^z = e^x(\cos y + i \sin y)$, so[1]

(77) $$|e^z| = e^x.$$

A complex function f of a complex variable is *analytic* on an open half-plane H iff it is differentiable at every point of H.

(78) Theorem.[2] *Suppose f is analytic on an open half-plane H. Then f has derivatives $f^{(n)}$ of all orders n at all points of H. If $z_0 \in H$, choose a positive real number r so small that*

$$D_r = \{z : |z - z_0| \leqq r\} \subset H.$$

Then

$$f(z) = \Sigma_{n=0}^{\infty} f^{(n)}(z_0) \cdot (z - z_0)^n / n!.$$

The series converges absolutely and uniformly on D_r.

(79) Theorem.[3] *If f_n is an analytic function on a half-plane and $f_n \to f$ uniformly on compact sets, then f is analytic.*

(80) Theorem.[4] *If f is an analytic function on a half-plane, and f vanishes on a set which has a point of accumulation, then f is identically* 0.

[1] Reference: (Ahlfors, 1953, p. 47).
[2] Reference: (Ahlfors, 1953, pp. 96 and 142). Hard.
[3] Reference: (Ahlfors, 1953, Theorem 1 on p. 138). Hard.
[4] Reference: (Ahlfors, 1953, p. 102). Hard.

BIBLIOGRAPHY

LARS V. AHLFORS (1953; 2nd ed., 1965). *Complex Analysis*, McGraw-Hill, New York.

DAVID BLACKWELL (1954). On a class of probability spaces, *Proc. 3rd Berk. Symp.*, Vol. 2, pp. 1–6.

DAVID BLACKWELL (1955). On transient Markov processes with a countable number of states and stationary transition probabilities, *Ann. Math. Statist.*, Vol. 26, pp. 654–658.

DAVID BLACKWELL (1958). Another countable Markov process with only instantaneous states, *Ann. Math. Statist.*, Vol. 29, pp. 313–316.

DAVID BLACKWELL (1962). Representation of nonnegative martingales on transient Markov chains, Mimeograph, Statistics Department, University of California at Berkeley.

DAVID BLACKWELL and LESTER DUBINS (1963). A converse to the dominated convergence theorem, *Illinois J. Math.*, Vol. 7, pp. 508–514.

DAVID BLACKWELL and DAVID A. FREEDMAN (1964). The tail σ-field of a Markov chain and a theorem of Orey, *Ann. Math. Statist.*, Vol. 35, pp. 1291–1295.

DAVID BLACKWELL and DAVID FREEDMAN (1968). On the local behavior of Markov transition probabilities, *Ann. Math. Statist.*, Vol. 39, pp. 2123–2127.

DAVID BLACKWELL and DAVID KENDALL (1964). The Martin boundary for Pólya's urn scheme and an application to stochastic population growth, *J. Appl. Probability*, Vol. 1, pp. 284–296.

R. M. BLUMENTHAL (1957). An extended Markov property, *Trans. Amer. Math. Soc.*, Vol. 85, pp. 52–72.

R. M. BLUMENTHAL and R. K. GETOOR (1968). *Markov Processes and Potential Theory*, Academic Press, New York.

R. M. BLUMENTHAL, R. GETOOR, and H. P. MCKEAN, Jr. (1962). Markov processes with identical hitting distributions, *Illinois J. Math.*, Vol. 6, pp. 402–420.

LEO BREIMAN (1968). *Probability*, Addison-Wesley, Reading.

D. BURKHOLDER (1962). Transient processes and a problem of Blackwell, Mimeograph, Statistics Department, University of California at Berkeley.

D. BURKHOLDER (1962). Successive conditional expectations of an integrable function, *Ann. Math. Statist.*, Vol. 33, pp. 887–893.

KAI LAI CHUNG (1960; 2nd ed., 1967). *Markov Chains with Stationary Transition Probabilities*, Springer, Berlin.

KAI LAI CHUNG (1963). On the boundary theory for Markov chains, I, *Acta. Math.*, Vol. 110, pp. 19–77.

KAI LAI CHUNG (1966). On the boundary theory for Markov chains, II, *Acta. Math.*, Vol. 115, pp. 111–163.

KAI LAI CHUNG and W. H. J. FUCHS (1951). On the distribution of values of sums of random variables, *Mem. Amer. Math. Soc.*, no. 6.

R. COGBURN and H. G. TUCKER (1961). A limit theorem for a function of the increments of a decomposable process, *Trans. Amer. Math. Soc.*, Vol. 99, pp. 278–284.

HARALD CRAMÉR (1957). *Mathematical Methods of Statistics*, Princeton University Press.

ABRAHAM DE MOIVRE (1718). *The Doctrine of Chances*, Pearson, London. Chelsea, New York (1967).

C. DERMAN (1954). A solution to a set of fundamental equations in Markov chains, *Proc. Amer. Math. Soc.*, Vol. 5, pp. 332–334.

W. DOEBLIN (1938). Sur deux problèmes de M. Kolmogorov concernant les chaînes dénombrables, *Bull. Soc. Math. France*, Vol. 66, pp. 210–220.

W. DOEBLIN (1939). Sur certains mouvements aléatoires discontinus, *Skand. Akt.*, Vol. 22, pp. 211–222.

MONROE D. DONSKER (1951). An invariance principle for certain probability limit theorems, *Mem. Amer. Math. Soc.*, no. 6.

J. L. DOOB (1942). Topics in the theory of Markoff chains, *Trans. Amer. Math. Soc.*, Vol. 52, pp. 37–64.

J. L. DOOB (1945). Markoff chains-denumerable case, *Trans. Amer. Math. Soc.*, Vol. 58, pp. 455–473.

J. L. DOOB (1953). *Stochastic Processes*, Wiley, New York.

J. L. DOOB (1959). Discrete potential theory and boundaries, *J. Math. Mech.*, Vol. 8, pp. 433–458, 993.

J. L. DOOB (1968). Compactification of the discrete state space of a Markov process, *Z. Wahrscheinlichkeitstheorie*, Vol. 10, pp. 236–251.

LESTER E. DUBINS and DAVID A. FREEDMAN (1964). Measurable sets of measures, *Pac. J. Math.*, Vol. 14, pp. 1211–1222.

LESTER E. DUBINS and DAVID A. FREEDMAN (1965). A sharper form of the Borel-Cantelli lemma and the strong law, *Ann. Math. Statist.*, Vol. 36, pp. 800–807.

LESTER E. DUBINS and GIDEON SCHWARZ (1965). On continuous martingales, *Proc. Nat. Acad. Sci. USA*, Vol. 53, pp. 913–916.

NELSON DUNFORD and JACOB T. SCHWARTZ (1958). *Linear operators, Part I*, Wiley, New York.

A. DVORETZKY, P. ERDÖS, and S. KAKUTANI (1960). Nonincrease everywhere of the Brownian motion process, *Proc. 4th Berk. Symp.*, Vol. 2, pp. 103–116.

E. B. DYNKIN (1965). *Markov Processes*, Springer, Berlin.

P. ERDÖS and M. KAC (1946). On certain limit theorems of the theory of probability, *Bull. Amer. Math. Soc.*, Vol. 52, pp. 292–302.

WILLIAM FELLER (1945). On the integro-differential equations of purely discontinuous Markoff processes, *Trans. Amer. Math. Soc.*, Vol. 48, pp. 488–515.

WILLIAM FELLER (1956). Boundaries induced by nonnegative matrices, *Trans. Amer. Math. Soc.*, Vol. 83, pp. 19–54.

WILLIAM FELLER (1957). On boundaries and lateral conditions for the Kolmogoroff differential equations, *Ann. of Math.*, Vol. 65, pp. 527–570.

WILLIAM FELLER (1959). Non-Markovian processes with the semigroup property, *Ann. Math. Statist.*, Vol. 30, pp. 1252–1253.

WILLIAM FELLER (1961). A simple proof for renewal theorems, *Comm. Pure Appl. Math.*, Vol. 14, pp. 285–293.

WILLIAM FELLER (1966). *An introduction to probability theory and its applications, Vol. 2*, Wiley, New York.

WILLIAM FELLER (1968). *An introduction to probability theory and its applications, Vol. 1, 3rd ed.*, Wiley, New York.

WILLIAM FELLER and H. P. MCKEAN, Jr. (1956). A diffusion equivalent to a countable Markov chain, *Proc. Nat. Acad. Sci. USA*, Vol. 42, pp. 351–354.

R. GETOOR (1965). Additive functionals and excessive functions, *Ann. Math. Statist.*, Vol. 36, pp. 409–423.

G. H. HARDY, J. E. LITTLEWOOD, and G. POLYA (1934). *Inequalities*, Cambridge University Press.

T. E. HARRIS (1952). First passage and recurrence distributions, *Trans. Amer. Math. Soc.*, Vol. 73, pp. 471–486.

T. E. HARRIS and H. ROBBINS (1953). Ergodic theory of Markov chains admitting an infinite invariant measure, *Proc. Nat. Acad. Sci. USA*, Vol. 39, pp. 860–864.

P. HARTMAN and A. WINTNER (1941). On the law of the iterated logarithm, *Amer. J. Math.*, Vol. 63, pp. 169–176.

FELIX HAUSDORFF (1957). *Set Theory*, Chelsea, New York.

EDWIN HEWITT and L. J. SAVAGE (1955). Symmetric measures on Cartesian products, *Trans. Amer. Math. Soc.*, Vol. 80, pp. 470–501.

E. HEWITT and K. STROMBERG (1965). *Real and Abstract Analysis*, Springer, Berlin.

G. A. HUNT (1956). Some theorems concerning Brownian motion, *Trans. Amer. Math. Soc.*, Vol. 81, pp. 294–319.

G. A. HUNT (1957). Markoff processes and potentials, 1, 2, 3, *Illinois J. Math.*, Vol. 1, pp. 44–93; Vol. 1, pp. 316–369; Vol. 2, pp. 151–213 (1958).

G. A. HUNT (1960). Markoff chains and Martin boundaries, *Illinois J. Math.*, Vol. 4, pp. 313–340.

K. ITÔ and H. P. MCKEAN, Jr. (1965). *Diffusion Processes and Their Sample Paths*, Springer, Berlin.

W. B. JURKAT (1960). On the analytic structure of semigroups of positive matrices, *Math. Zeit.*, Vol. 73, pp. 346–365.

A. A. JUSKEVIC (1959). Differentiability of transition probabilities of a homogeneous Markov process with countably many states, *Moskov. Gos. Univ. Učenye Zapiski*, No. 186, pp. 141–159; in Russian. Reviewed in *Math. Rev.* No. 3124 (1963).

M. KAC (1947). On the notion of recurrence in discrete stochastic processes, *Bull. Amer. Math. Soc.*, Vol. 53, pp. 1002–1010.

S. KAKUTANI (1943). Induced measure preserving transformations, *Proc. Impl. Acad. Tokyo*, Vol. 19, pp. 635–641.

J. G. KEMENY and J. L. SNELL (1960). *Finite Markov Chains*, Van Nostrand, Princeton.

J. G. KEMENY, J. SNELL, and A. W. KNAPP (1966). *Denumerable Markov Chains*, Van Nostrand, Princeton.

A. KHINTCHINE (1924). Ein Satz der Wahrscheinlichkeitsrechnung, *Fund. Math.*, Vol. 6, pp. 9–20.

J. F. C. KINGMAN (1962). The imbedding problem for finite Markov chains, *Z. Wahrscheinlichkeitstheorie*, Vol. 1, pp. 14–24.

J. F. C. KINGMAN (1964). The stochastic theory of regenerative events, *Z. Wahrscheinlichkeitstheorie*, Vol. 2, pp. 180–224.

J. F. C. KINGMAN (1968). On measurable p-functions, *Z. Wahrscheinlichkeitstheorie*, Vol. 11, pp. 1–8.

J. F. C. KINGMAN and STEVEN OREY (1964). Ratio limit theorems for Markov chains, *Proc. Amer. Math. Soc.*, Vol. 15, pp. 907–910.

FRANK KNIGHT and STEVEN OREY (1964). Construction of a Markov process from hitting probabilities, *J. Math. Mech.*, Vol. 13, pp. 857–873.

A. KOLMOGOROV (1931). Über die analytischen Methoden in der Wahrscheinlichkeitsrechnung, *Math. Ann.*, Vol. 104, pp. 415–458.

A. KOLMOGOROV (1936). Anfangsgründe der Theorie der Markoffschen Ketten mit unendlichen vielen möglichen Züstanden, *Mat. Sb.*, pp. 607–610.

A. KOLMOGOROV (1951). On the differentiability of the transition probabilities in stationary Markov processes with a denumerable number of states, *Moskov. Gos. Univ. Učenye Zapiski*, Vol. 148, Mat. 4, pp. 53–59; in Russian. Reviewed on p. 295 of *Math. Rev.* (1953).

ULRICH KRENGEL (1966). On the global limit behavior of Markov chains and of general nonsingular Markov processes, *Z. Wahrscheinlichkeitstheorie*, Vol. 4, pp. 302–316.

CASIMIR KURATOWSKI (1958). *Topologie I*, 4th ed. Warsaw.

PAUL LÉVY (1951). Systèmes markoviens et stationnaires. Cas dénombrable, *Ann. Sci. École Norm. Sup.*, (3), Vol. 68, pp. 327–381.

PAUL LÉVY (1952). Complément à l'étude des processus de Markoff, *Ann. Sci. École Norm. Sup.*, (3) Vol. 69, pp. 203–212.

PAUL LÉVY (1953). Processus markoviens et stationnaires du cinquième type, *C. R. Acad. Sci. Paris*, Vol. 236, pp. 1630–1632.

PAUL LÉVY (1954). *Le Mouvement Brownien*, Gauthier Villars, Paris.

PAUL LÉVY (1954a). *Théorie de l'Addition des Variables Aléatoires*, Gauthier Villars, Paris.

PAUL LÉVY (1958). Processus markoviens et stationnaires. Cas dénombrable, *Ann. Inst. H. Poincaré*, Vol. 16, pp. 7–25.

PAUL LÉVY (1965). *Processus Stochastiques et Mouvement Brownien*, Gauthier Villars, Paris.

MICHEL LOÈVE (1963). *Probability Theory*, 3rd ed., Van Nostrand, Princeton.

A. A. MARKOV (1906). Extension of the law of large numbers to dependent events, *Bull. Soc. Phys. Math. Kazan.*, (2), Vol. 15, pp. 135–156; in Russian.

JACQUES NEVEU (1965). *Mathematical Foundations of the Calculus of Probability*, Holden-Day, San Francisco.

STEVEN OREY (1962). An ergodic theorem for Markov chains, *Z. Wahrscheinlichkeitstheorie*, Vol. 1, pp. 174–176.

DONALD ORNSTEIN (1960). The differentiability of transition functions, *Bull. Amer. Math. Soc.*, Vol. 66, pp. 36–39.

DANIEL RAY (1956). Stationary Markov processes with continuous paths, *Trans. Amer. Math. Soc.*, Vol. 82, pp. 452–493.

DANIEL RAY (1967). Some local properties of Markov processes, *Proc. 5th Berk. Symp.*, Vol. 2, part 2, pp. 201–212.

G. E. H. REUTER (1957). Denumerable Markov processes and the associated contraction semigroups on ρ, *Acta Math.*, Vol. 97, pp. 1–46.

G. E. H. REUTER (1959). Denumerable Markov processes. *J. London Math. Soc.*, Vol. 34, pp. 81–91.

G. E. H. REUTER (1969). Remarks on a Markov chain example of Kolmogorov, *Z. Wahrscheinlichkeitstheorie*, Vol. 13, pp. 315–320.

F. RIESZ and B. SZ. NAGY (1955). *Functional Analysis*, Ungar, New York.

B. A. ROGOZIN (1961). On an estimate of the concentration function, *Theor. Probability Appl.*, Vol. 6, pp. 94–96.

H. L. ROYDEN (1963). *Real Analysis*, Macmillan, New York.

STANISLAW SAKS (1964). *Theory of the Integral*, 2nd rev. ed., Dover, New York.

JAMES SERRIN and D. E. VARBERG (1969). A general chain rule for derivatives and the change of variables formula for the Lebesgue integral, *Amer. Math. Monthly*, Vol. 76, pp. 514–520.

A. SKOROKHOD (1965). *Studies in the Theory of Random Processes*, Addison-Wesley, Reading.

GERALD SMITH (1964). Instantaneous states of Markov processes, *Trans. Amer. Math. Soc.*, Vol. 110, pp. 185–195.

J. M. O. SPEAKMAN (1967). Two Markov chains with a common skeleton, *Z. Wahrscheinlichkeitstheorie*, Vol. 7, p. 224.

FRANK SPITZER (1956). A combinatorial lemma and its applications to probability theory, *Trans. Amer. Math. Soc.*, Vol. 82, pp. 323–339.

FRANK SPITZER (1964). *Principles of Random Walk*, Van Nostrand, Princeton.

VOLKER STRASSEN (1964). An invariance principle for the law of the iterated logarithm, *Z. Wahrscheinlichkeitstheorie*, Vol. 3, pp. 211–226.

VOLKER STRASSEN (1966). A converse to the law of the interated logarithm, *Z. Wahrscheinlichkeitstheorie*, Vol. 4, pp. 265–268.

VOLKER STRASSEN (1966a). Almost sure behavior of sums of independent random variables and martingales, *Proc. 5th Berk. Symp.*, Vol. 2, part 1, pp. 315–343.

H. F. TROTTER (1958). A property of Brownian motion paths, *Illinois J. Math.*, Vol. 2, pp. 425–433.

A. WALD (1944). On cumulative sums of random variables, *Ann. Math. Statist.*, Vol. 15, pp. 283–296.

N. WIENER (1923). Differential space, *J. Math. and Phys.*, Vol. 2, pp. 131–174.

DAVID WILLIAMS (1964). On the construction problem for Markov chains, *Z. Wahrscheinlichkeitstheorie*, Vol. 3, pp. 227–246.

DAVID WILLIAMS (1966). A new method of approximation in Markov chain theory and its application to some problems in the theory of random time substitution, *Proc. Lond. Math. Soc. (3)*, Vol. 16, pp. 213–240.

DAVID WILLIAMS (1967). Local time at fictitious states, *Bull. Amer. Math. Soc.*, Vol. 73, pp. 542–544.

DAVID WILLIAMS (1967a). A note on the Q-matrices of Markov chains, *Z. Wahrscheinlichkeitstheorie*, Vol. 7, pp. 116–121.

David Williams (1967b). On local time for Markov chains, *Bull. Amer. Math. Soc.*, Vol. 73, pp. 432–433.

Helen Wittenberg (1964). Limiting distributions of random sums of independent random variables, *Z. Wahrscheinlichkeitstheorie*, Vol. 1, pp. 7–18.

A. Zygmund (1959). *Trigonometric Series*, Cambridge University Press.

Additional references

C. Dellacherie and P. A. Meyer (1978). *Probabilités et potentiel*, Hermann, Paris.

F. B. Knight (1981). Essentials of Brownian motion and diffusion, *Math. Surveys*, Vol. 18.

H. P. McKean (1969). *Stochastic Integrals*, Academic Press, New York.

David Williams (1979). *Diffusions, Markov. Processes, and Martingales*, Wiley, New York.

INDEX

SYMBOL FINDER

I've listed here the symbols with some degree of permanence. The listing is alphabetical, first English then Greek. Where possible, there is a quick definition and a page reference. The list is partial, and local usage is sometimes different. You may wish to look at Sections 3.1-3, on notation and references.

ENGLISH

$\mathscr{A}(t)$: admissible σ-field, page 21
$\mathscr{A}(\tau+)$: pre-τ sigma-field, page 21
B: normalized Brownian motion, page 1
$BC(I)$: bounded, continuous functions on I, pages 104–105
B_x: standard Brownian motion, page 103
$\mathscr{B}$: Borel σ-field in Ω, page 102
$\mathscr{B}(t)$: σ-field to time t, page 102
$\mathscr{B}(t+)$: pre-t sigma-field, page 102
$\mathscr{B}(\tau+)$: pre-τ sigma-field, page 102
$C(I)$: continuous functions on I, page 14
$C[0, \infty)$: continuous functions on $[0, \infty)$
$C[0, 1]$: continuous functions on $[0, 1]$, pages, 4, 49
$C[0, 1]^\infty$: sequences in $C[0, 1]$, page 51
$d(f, g) = \|f - g\|$, page 49
$d(f, K) = \inf\{d(f, g): g \in K\}$, page 51
$e_{ab}(x) = e_J(x) = E_x\{\tau_{ab}\}$, pages 104, 126
$e_b(x) = E_x\{\tau_b\}$, page 135
$e(x) = E_0\{\tau_x\}$, page 136
$\exp x = e^x$
E_x is P_x-expectation, Sections 2.1-7
E_x is B_x-expectation, Sections 2.8-9

GREEK

π_σ: distribution of Brownian motion with variance σ^2 and time in [0, 1], page 77
Σ: Borel σ-field in $C[0, 1]$, page 49
τ: Markov time, pages 1, 21, 102
τ_a: time to hit a, page 108
τ_{ab}: time to hit a or b, pages 104, 111
$\varphi = \sup\{t: M(t) < \infty\}$, page 149
$\omega \in \Omega = \Omega_I$: continuous functions from $[0, \infty)$ to I, page 102
Ω_m: set, page 150